Motivated mathematics

Motivated mathematics

A. EVYATAR & P. ROSENBLOOM

CAMBRIDGE UNIVERSITY PRESS

Cambridge

London   *New York*   *New Rochelle*

Melbourne   *Sydney*

CAMBRIDGE UNIVERSITY PRESS
Cambridge, New York, Melbourne, Madrid, Cape Town, Singapore, São Paulo, Delhi

Cambridge University Press
The Edinburgh Building, Cambridge CB2 8RU, UK

Published in the United States of America by Cambridge University Press, New York

www.cambridge.org
Information on this title: www.cambridge.org/9780521233088

© Cambridge University Press 1981

First published 1981
Re-issued in this digitally printed version 2008

A catalogue record for this publication is available from the British Library

ISBN 978-0-521-23308-8 paperback

CONTENTS

PREFACE

Many distinguished mathematicians and educators, going back at least to F. Klein and E. H. Moore about 80–90 years ago, have advocated that mathematics should be taught in close connection with its applications and be motivated by them. The critics of the reforms in the United States' secondary school mathematics curriculum in the 1950s strongly emphasized this view, but produced very little concrete material for school mathematics showing how mathematics can be applied. Up to that time the only applications which were treated in school mathematics were consumers' and shopkeepers' problems.

In 1963, Professor Paul Rosenbloom, then Director of the Minnesota School Mathematics and Science Center, produced a first draft of a course on computer and applied mathematics for the twelfth grade. The mathematical topics were chosen in accordance with the recommendations of the Commission on Mathematics of the College Entrance Examination Board (C.E.E.B.). However, the treatment of topics in the course was original, and different from, for example that of the School Mathematics Study Group; each topic was motivated by beginning with a concrete application to the natural or social sciences whose investigations would lead to that topic. A source of inspiration for the course was the book by Th. von Karman and M. Biot, *Mathematical methods in engineering* (McGraw-Hill, New York, 1940).

For several years, beginning in 1963, summer institutes for high school teachers and students were conducted at the University of Minnesota. The two groups were taught together so that the teachers could observe the students learning. Besides the material on applied mathematics, the two groups were given courses in computer programming and numerical analysis, and a seminar in problem solving.

During the following school year the participating teachers taught the material on applied mathematics to their twelfth grade classes. Computer manufacturers (Control Data and Univac) in the Twin Cities area provided these classes with computer time. Mathematicians at the University of Minnesota met with the participating teachers once a week to discuss the problems they encountered. For the first two summers only the best high school applicants were accepted, but thereafter the high school applicants were selected randomly so that the teachers could observe the reactions of typical classes to the materials. Professor Rosenbloom started to teach this material himself to several classes of graduate students in mathematical education when he moved to Teachers College.

Professor Evyatar has been teaching pre- and in-service courses for mathematics teachers at the Technion, Israel Institute of Technology, for about 20 years, and he has also written textbooks for use at the secondary school level.

The collaboration of the authors began during the academic year 1976–7 when Professor Evyatar was a Visiting Professor at Teachers College. At that time the existing material was combined and revised, and much new material was added. During 1978-9, when Professor Rosenbloom visited the Technion, the material was again revised and amended. The present version is intended specifically for the pre- and in-service education of secondary school mathematics teachers.

Our thanks to Rick Troxel for his comments and to the staff of the Cambridge University Press, particularly Mrs J. Holland, for their efficient help.

A.E.
P.R.

INTRODUCTION

The mathematical content of this book takes into account the recommendations of the Commission on Mathematics of the College Entrance Examination Board (C.E.E.B.) as applied to the U.S.A., European countries, and
Japan.

The C.E.E.B. recommends that high schools should provide courses on
elementary functions, probability, statistics, analytical geometry, calculus,
and linear algebra, all of which should be available as one-semester courses.
They also suggest some discussion of limits.

Syllabi for secondary school examinations conducted by the Oxford and
Cambridge School Examination Boards differ from the above list of topics
mainly by the emphasis that is put on limits and calculus, and by the extent
of the knowledge required in probability and statistics. Analytical geometry
and calculus are standard topics in the secondary school curriculum in Israel,
Japan and many European countries.

The following list shows which chapters in this book are related to a specific
mathematical topic:

Polynomials	1
Logarithmic functions	2
Exponential functions	3
Trigonometric functions	7
Limits	3, 6
Probability and statistics	4
Analytical geometry and calculus	1, 2, 3, 6, 8
Linear algebra	8

It is not expected that an average high school teacher will be able to teach
an entire course in applied mathematics until adequate materials are available
for school use. We believe, however, that a teacher who has studied this book
will be able to introduce some of the topics here, where appropriate, to

motivate and illustrate the courses he is already teaching. We have often indicated how some aspects of these topics can be taught at an even earlier stage. It is valuable for the secondary or high school teacher to see how elementary many of the essential ideas are.

There are two fundamental problems in the teaching of applications. One is that the teacher must learn something about the field to which the mathematics is applied as well as the mathematics. This difficulty has limited us to topics in which the non-mathematical material could be explained very briefly.

The second difficulty is that true applications of mathematics involve showing how the mathematical formulation arises from a problem in the real world and how the mathematical results are to be interpreted. The common practice of merely stating mathematical problems in 'applied' language does not teach anything valuable about applied mathematics. Sometimes carrying the analysis far enough to obtain a really applicable result requires mathematics beyond the scope of this course. In such cases, we have tried to illustrate the ideas by obtaining at least some non-trivial results with interesting concrete interpretations.

For the users of this material the main objective is to teach mathematics, and the applications serve mainly for motivation purposes. We have therefore tried to show how much of the drill in mathematical techniques can be incorporated in a natural way in the course of analyzing the mathematical models studied. We thus have an amusing inversion of aims: physics, biology, linguistics, etc., are here applied to the teaching of mathematics.

Most of the exercises in the present text are intended to lead the student to take an active part in the development of the subject. The students' work should be used as the basis of class discussions and subsequent lessons should make considerable use of the students' results. We have included comparatively few routine exercises, since an adequate supply of such problems is available in standard texts. The teacher may wish to supplement our exercises with such material.

The organization of the text is designed to facilitate its use as an actual textbook or as a random-access reference work for teachers. Each chapter treats a specific mathematical notion – be it logarithms or vectors or matrices – at different levels of ability. The main sections are at the secondary school or high school level, but in most of the chapters there are also curricular units that can be used by teachers of grades 5–8 (ages 11–14), and sections for more advanced students. These more difficult sections serve a dual purpose: they give a view of topics dealt with at the secondary school level from a higher standpoint, and also provide a text suitable for teaching at the college level.

Each chapter has an introductory section which explains the content and level of treatment of each of the subsequent sections. The following table classifies the different parts of each chapter according to the mathematical sophistication of their content. 'Elementary' sections can be taught to ages 11-14 (grades 5-8). The 'intermediate' sections are for high school students (ages 14-18), and 'advanced' sections are beyond the intermediate level.

This is, of course, only our own view of the relative difficulty of the treatment, and every teacher should best judge for himself which material he can use and at what time.

Table showing mathematical content for each chapter, with level of treatment by section

Chapter	Mathematical content	Level of treatment		
		elementary	intermediate	advanced
1	Computing and polynomials	1.1,1.2	1.3–1.6	
2	Logarithmic functions	2.4	2.1,2.2,2.5	2.6
3	Exponential functions	3.3	3.1,3.2	3.4
4	Statistics	4.2	4.1,4.3,4.4	4.5
5	Optimalization	5.1,5.3(part)	5.2,5.3(part),5.4	5.5
6	Limits	6.1	6.2,6.3	6.4
7	Trigonometric functions	7.1	7.2,7.3	7.4
8	Linear transformations	8.2	8.1	8.3

1

Calculators and programming: applications to polynomials

A good deal of primary data in mathematics consists of tables and graphs. These data make the abstract formulas and equations concrete and down to earth. They also exhibit phenomena which stimulate investigation. Much of the theory can be motivated by the desire to explain what is observed, or to answer natural questions arising out of the data. Most mathematical theories have originated in the search for answers to specific questions. To present the theory divorced from the data is to give the students answers to questions they have never asked.

The data usually require computation. The main purpose of the calculation is to find which phenomena the data exhibit. The drudgery of the computation and concern over computational error are distractions from this purpose. Therefore the students should be encouraged to use computational aids such as hand or desk calculators. Many schools and colleges also have computer facilities available. The most common language used in educational computers is BASIC.

We shall give here an introduction to the use of calculators and to BASIC programming, to be used in conjunction with the later chapters. There are minor variations in the versions of BASIC used in different installations, so that our explanations may have to be modified slightly in some institutions.

1.1   The calculator: learning how to use it

We use the term calculator for any hand-held calculator. Obviously, in any case, one should first make sure that all students know how to use it. This entails some care on the part of the instructor, since he should check that, although different makes or models may be used concurrently, all students can perform the same operations.

A reasonable procedure would be to go over the following points:

(*a*)   Read the manual of instructions. This explains what operations can be performed on the calculator, and how to do them. Make up a list of the possibilities.

(b) Practice with some simple calculations first, especially problems whose answers you already know.

(c) How does the calculator handle decimal points? Try problems like

$1/10, 1/100, \ldots$

and see what you get. Does it handle negative numbers? Try

$0 - 10$, etc.

(d) What about large numbers? Try

$10 \times 100, 10 \times 1000, \ldots$

How does it round off?

(e) How does it respond to division by zero?

(f) How does it handle parentheses? How do you compute

$(2 + 3) \times 4$ or $2 + (3 \times 4)$

with your calculator?

After this elementary work, we go over to learning how our calculator handles functions. It is important to insist on a unified way of writing down the order of hitting the keys, for instance

$$\boxed{4} \quad \boxed{y^x} \quad \boxed{6} \quad \boxed{=} \quad 4096 \quad,$$

$$\boxed{1024} \quad \boxed{\sqrt{x}} \quad \boxed{=} \quad 32 \quad.$$

(Keys are denoted by $\boxed{}$; the display is not indicated. The calculator used here is a T.I. 30.)

We can now solve more complicated problems:

(a) Solve the equation

$3.14x + 2.27 = 51.3x - 79.6$.

The solution looks like this:

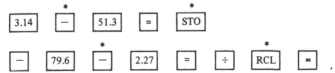

Explain the starred keys. How would we write this for

$ax + b = cx + d$?

When do we use $\boxed{+/-}$?

(b) Use the calculator for theoretical work. For example, to show the distributive law as an efficiency device, compute

$(3 \times 6) + (5 \times 6) + (7 \times 6)$.

The key-hitting series looks like

$\boxed{3}\ \boxed{\times}\ \boxed{6}\ \boxed{+}\ \boxed{5}\ \boxed{\times}\ \boxed{6}\ \boxed{+}\ \boxed{7}\ \boxed{\times}\ \boxed{6}\ \boxed{=}$,

altogether 12 key-strokes. What happened to the parentheses? Another way of obtaining the same result is to compute

$(3 + 5 + 7) \times 6$

or

$\boxed{3}\ \boxed{+}\ \boxed{5}\ \boxed{+}\ \boxed{7}\ \boxed{=}\ \boxed{\times}\ \boxed{6}\ \boxed{=}$,

altogether 9 key-strokes. Why did we insert $\boxed{=}$ between $\boxed{7}$ and $\boxed{\times}$? Do we always save key-strokes by using the distributive principle?

(c) Consider the problem of computing a_n, where

$a_1 = 3,$

$a_n = 3a_{n-1}.$

(This comes up on p. 28.)

In considering (c), you want to compute a_n for a given value of n. You might make tally marks to keep track of when you have reached the given n. The work could be planned schematically as in fig. 1.1, in which case you

Fig. 1.1

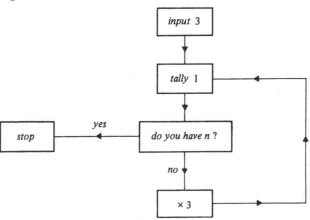

punch 3 into the calculator, make a tally on your record, and compare with n. If you are already at n, you stop. If not, you multiply by 3, then go back to the second step. You repeat until you reach n tally marks, then you stop.

This plan of steps is called a *flow-chart*. It is useful to make a flow-chart if the problem is at all complicated. A flow-chart is also helpful in programming for a computer.

In the above case, a_n is, of course, 3^n. If your calculator can raise numbers to powers you can calculate a_n in one step and compare results. For large n, the calculator may round off differently in the two procedures. Try it and see.

1.2 Elementary programming

To program the same problem in BASIC, we would want the computer to do the tallying, so our first step is to give the input n. This will store the number at a certain place in the memory which is now labeled n. (You can use any letter of the alphabet as a variable in this way.) Next you want to insert 3 in the a-place in the memory.

The simplest way to do the tallying on the computer is to subtract 1 from n repeatedly until you reach 1. This operation of subtraction can be described as:

Calculate $n - 1$ and put the result in the n-place of the memory,

or more simply like this:

$$n - 1 \to n.$$

In BASIC this command is written

LET N = N − 1;

the computer will only print capital letters so the variables a and n become A and N in BASIC. The equality sign here is used differently from the way it is used ordinarily in mathematics. You may interpret it as a reversed arrow ($\leftarrow$), and you may describe it as follows:

Put in the n-place (replacing if necessary what is already there) the number $n - 1$, calculated from the present state of the memory.

Similarly the next step in making the flow-chart would be, in BASIC, the command

LET A = A * 3

(* is the multiplication sign in BASIC). The flow-chart for our program is shown in fig. 1.2. We have added a command to print the answer.

In writing the program, we must number the commands:

```
10      INPUT N
20      LET A = 3
30      IF N = 1 THEN 70
40      LET A = A * 3
50      LET N = N − 1
60      GØ TØ 30
70      PRINT A
80      END   .
```

Fig. 1.2.

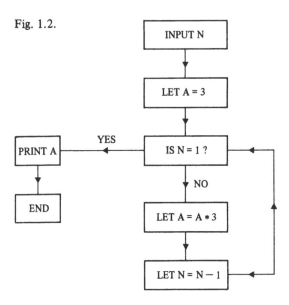

In writing this program, we left the space after 'THEN' in the third command blank until we reached the 'GØ TØ' command. Then we could see that the alternative should come after command 60, and write '70' in the third line. The computer usually executes the commands in numerical order. One obvious exception is a 'GØ TØ' command. With an 'IF ... THEN ...' command, if the answer is 'NØ' the computer goes to the next command, but if it is 'YES' then the computer goes where the 'THEN' tells it.

If you type in this program, and then want to run it, you simply type the command

RUN

The computer will respond with a question mark:

?

You then type the value of *n*. The computer will then print the answer, which will be a_n.

If we want the computer to print a table of a_m for $1 \leqslant m \leqslant n$, we must modify the above procedure. It is convenient to start with $m = 1$ and to add 1 repeatedly until we reach *n*. The flow-chart in fig. 1.3 describes the method. We can now write the program:

```
10    INPUT N
20    PRINT "M", "A"
30    LET M = 1
```

```
 40     LET A = 3
 50     PRINT M, A
 60     IF M = N THEN 100
 70     LET M = M + 1
 80     LET A = A * 3
 90     GØ TØ 50
100     END   .
```

Note that in command 20 the computer will print exactly what is enclosed in the quotation marks, namely the letters 'M' and 'A'. The comma in the command ensures that there will be a reasonable space between them on the print-out. Command 50 tells the computer to print what is in the *m*- and *a*-places in

Fig. 1.3.

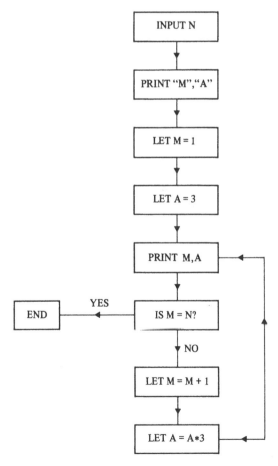

the memory. Since we have a comma there also, the numbers will be printed under the proper headings.

As we will want to refer to the results later, we should label them to know what they represent. This entails adding the following commands:

     13   PRINT "TABLE OF A(M)"
     16   PRINT "A(M + 1) = 3 * A(M)"
     18   PRINT .

In BASIC all symbols are written on a line, with no subscripts, superscripts, or exponents. Thus we write 'A(M)' for a_m, 'A(M + 1)' for $a_{(m + 1)}$. Since the computer executes commands in *numerical* order, it will perform these between commands 10 and 20. This is why it is advantageous to have an interval between command numbers, so that it will be easy to modify the program. Command 18 merely leaves an empty line, for appearance's sake. Here is a typical print-out:

     ? 5
     TABLE OF A(M)
     A(M + 1) = 3 * A(M)
     M   A
     1   3
     2   9
     3   27
     4   81
     5   243 .

Try it yourself. Incidentally, if you run this program with the input 20, you will find out how your computer handles roundoff.

1.3   Programming: the next stage

Suppose that you need, in addition to the above print-out, a table of the values of a_m for $1 \leqslant m \leqslant n$, when a_m satisfies the equations

$$a_1 = 7,$$
$$a_m = 7a_{m-1}.$$

To achieve this you would merely have to replace the '3' by '7' in commands 16, 40, and 80 of the above program.

For this purpose it is not necessary to re-type the whole program. It is sufficient to type these new commands:

     16   PRINT "A(M + 1) = 7 * A(M)"
     40   LET A = 7
     80   LET A = A * 7.

The computer will simply replace the old commands with these numbers by the new commands. You can carry out corrections to your program in the same way.

If you have made some additions and changes in your program, you may wish to see what program is now stored in the computer. You need merely type the command

    LIST

and the computer will list the current program with the commands in *numerical order*.

Other things being equal, it is better to make your program *general* in the first place, so that you can use it for a whole family of problems by simply changing the inputs. Thus we should prefer to modify the commands as follows:

    10   INPUT N, B
    16   PRINT "A(M + 1) = B * A(M)"
    19   PRINT "B = "B
    40   LET A = B
    80   LET A = A * B .

When you run the program and the computer prints '?', you will supply *two* numbers, which will be the values of n and b respectively. What happens if command 19 is given as 'PRINT "B = B" '?

On pp. 48–50, chapter 2, you often need to find m such that

$$b^m \leqslant c < b^{m+1},$$

where b and c are given numbers greater than 1, and then to compute c/b^m. A flow-chart for this computation might be as shown in fig. 1.4. In this figure we have added steps to print the answer and also the value of the quotient c/a, which is needed for the next part of the work.

Write a program for this process together with a suitable heading. How many commands are there in the program?

This problem illustrates the characteristic power of the digital computer. The loop in the chart in fig. 1.4, which will be expressed by a 'GØ TØ' and an 'IF ... THEN...' command in the program, tells the computer to do certain operations repeatedly until a certain event occurs. In this way you can make the computer perform thousands of steps with a short program. For example, here if b is close to 1, say $b = 1.001$, and $c = 2$, then the computer will do about 1000 multiplications and about 2000 other simple operations. Run the program and see how long it takes. (Some computers will report the computation time.)

Typically, in many elementary problems it takes longer to feed the program

Fig. 1.4.

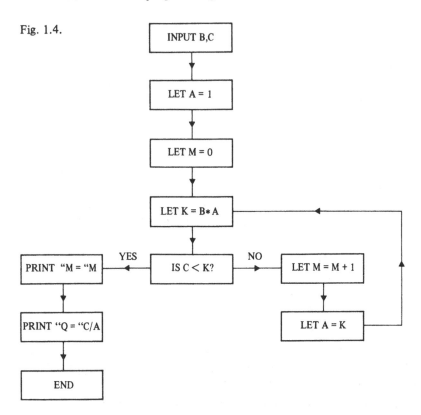

into the computer and for it to type out the answers than it does for the computer to calculate the answers.

The difference between the loop in this problem and the one in the previous problem is that here the repetition is carried out until a certain event (here $c < k$) occurs, and the number of repetitions is *not* known in advance, whereas in the previous problem we want the computer to repeat certain operations a *prescribed* number of times (there n times). The latter type of loop can also be programmed in another way which is often simpler. We can use instead the program:

```
30    INPUT N, B
40    PRINT "B = "B
50    PRINT
60    PRINT "M", "A"
70    LET A = B
80    FØR M = 1 TØ N
90    PRINT M, A
```

```
100     LET A = A * B
110     NEXT M
120     END  .
```

In this program, beginning with command 80, the computer sets $m = 1$, executes the commands until it reaches the 'NEXT M' command, replaces m by its next value, goes back to command 90, and repeats until m reaches the value n. Then the computer goes to the command after the 'NEXT M' command. In this simple problem the two types of program are about equal in length. In more complicated problems 'FØR...NEXT...' procedure is often shorter.

If you want the variable to proceed in steps of some other size than 1, you can simply use a command of the form

FØR M = C TØ D STEP S.

Then the computer executes the program for $m = c, c + s, c + 2s, \ldots$, until m reaches or passes d. For example, here is a program for computing a table of $x(t)$ from the equation

$$x[(n + 1)h] = (1 + rh)x(nh)$$

on p. 59, chapter 3, given the value of $x(0)$:

```
40      INPUT R, H, A, K
50      PRINT "R = "R, "X(0) = "A
60      PRINT
65      LET B = 1 + R * H
70      PRINT "T", "X"
80      LET X = A
90      FØR T = 0 TØ K STEP H
100     PRINT T, X
110     LET X = B * X
120     NEXT T
130     END  .
```

Try this program with particular values for the inputs.

If you want the variable to decrease in regular steps, you can use the same type of command with a *negative* value for s. If the size of s is not specified, the computer always takes it to be 1.

1.4   Calculators as motivation for theoretical work

As mentioned in section 1.1, it is important to remove drudgery from computation when teaching mathematics through its applications, as computation enables us to examine various mathematical models and assess their predictions. It is also obvious that, since 'round numbers' are irrelevant when working with a calculator, we can use realistic data. It is usually assumed that a calculator helps the understanding of limiting processes. This is in fact true – but with a slight twist. The characteristics of the calculator interfere with the limiting process.

Try, for example, to sum the infinite series

$$\sum_{n=1}^{\infty} \frac{1}{n}$$

by means of the computer. An obvious program would be

```
10      INPUT M
20      LET S = 0
30      FØR N = 1 TØ M
40      LET S = S + (1/N)
50      NEXT N
60      PRINT S  ,
```

which would give us

$$s_m = \sum_{n=1}^{m} \frac{1}{n} \, .$$

However, beyond a certain value of n the computer will round off $1/n$ to 0, and the value of s_m from that point on will be constant.

This is even more apparent on a calculator. Using some special tricks on a programmable calculator, Professor D. B. Scott of the University of Sussex was able to obtain a value of s_m (for a certain large m), which was close to 24, but he did not succeed in reaching 24.

Let us look at the problem a little more closely. Suppose we group the terms like this:

$$\frac{1}{1} + \left(\frac{1}{2} + \frac{1}{3} \right) + \left(\frac{1}{4} + \frac{1}{5} + \frac{1}{6} + \frac{1}{7} \right) + \dots \,,$$

beginning each group of terms with $1/2^k$ and taking 2^k terms in each group. In the third group, for example, the largest term is the first term, $1/4$, and there are four terms. We infer that

$$\frac{1}{4} + \frac{1}{5} + \frac{1}{6} + \frac{1}{7} < 4 \times \frac{1}{4} = 1.$$

In the same way we see that the sum of each group after the term $1/1$ is less than 1. We thus obtain

$$\frac{1}{1} + \frac{1}{2} + \frac{1}{3} + \ldots + \frac{1}{2^{k+1}-1} < k+1 \text{ for } k > 1.$$

Thus the sum of the first $2^{1\,000\,000} - 1$ terms is less than $1\,000\,000$.

Similarly each term in the third group is larger than the first term in the next group, $1/8$. Hence we find that

$$\frac{1}{4} + \frac{1}{5} + \frac{1}{6} + \frac{1}{7} > 4 \times \frac{1}{8} = \frac{1}{2}.$$

In the same way, we find that the sum of each group is greater than $1/2$. This yields the estimate

$$\frac{1}{1} + \frac{1}{2} + \frac{1}{3} + \ldots + \frac{1}{2^{k+1}-1} > \frac{k+1}{2}.$$

Thus if the values of the terms were computed exactly, we would find that $s_m > 24$ for $m = 2^{48} - 1$. This number is greater than 10^{14}, since $2^{10} > 10^3$ and $2^8 > 10^2$. If the computer could compute $1/n$ and add it to the previous sum (command 40 in the above program) in 10^{-6} seconds, it would take the computer more than 10^8 seconds to compute s_m for this value of m. Estimate the number of seconds in a year.

Exercises

1. Estimate how long the computer would take to reach a value of 60 in the above problem. Compare with the estimated age of the Earth.
2. Estimate how many terms are needed in the above problem to reach a sum greater than $1\,000\,000$.
3. Compute

$$\frac{(3+h)^2 - 3^2}{h}$$

for $h = 1/2^n$, $n = 1,2,3, \ldots$, using a hand or desk calculator or a computer. What happens? Why?

We see that the computer has several limitations in the study of limiting processes:

(a) the roundoff of the computer may disguise what really happens;

(b) an impractically long time may be needed to reveal the real trend.

Nevertheless, the computer or calculator can be a valuable tool in the investigation of limits. These instruments can often suggest trends and exhibit phenomena. The theoretical analysis often arises out of concrete numerical results and can frequently be motivated by them.

Exercises

4. Let $x_0 = 1$ and let

$$x_{n+1} = \frac{1}{1 + x_n} .$$

   Compute x_n for $n = 1, 2, 3, \ldots$, and also $x_n(x_n + 1)$. What seems to happen? What conjectures do your numerical results suggest?

5. Choose any values for x_0 and x_1. Compute x_n for $n \geqslant 2$ by

$$x_{n+2} = x_{n+1} + x_n,$$

   and compute x_{n+1}/x_n. What seems to happen? Let

$$y_n = x_{n+1}/x_n,$$

   and compute

$$y_n^2 - y_n.$$

6. Choose any two positive numbers a and b. Let

$$x_0 = a, y_0 = b,$$
$$x_{n+1} = (x_n y_n)^{\frac{1}{2}}, y_{n+1} = (x_n + y_n)/2.$$

   Compute x_n and y_n for $n = 1, 2, 3, \ldots$ What seems to happen?

7. Choose any three positive numbers, a, b, and c. Let

$$x_0 = a, y_0 = b, z_0 = c,$$
$$x_{n+1} = \frac{3}{1/x_n + 1/y_n + 1/z_n} , y_{n+1} = (x_n y_n z_n)^{\frac{1}{3}},$$

   and

$$z_{n+1} = \frac{x_n + y_n + z_n}{3} .$$

   Compute x_n, y_n, and z_n for $n = 1, 2, 3, \ldots$ What seems to happen?

1.5 Efficiency of programs: computation of polynomials

We have seen above that sometimes it is desirable to give some preliminary thought before engaging in computation or programming. If a certain process must be repeated many times, one may have to think about the time required. Even if the computer performs the individual operations very rapidly, the saving of a few steps in each repetition of the process may save a significant amount of computer time.

Let us illustrate this with the problem of computing a polynomial:

$$P(x) = ax^2 + bx + c.$$

We could, in BASIC, give the command

LET Y = (A * X * X) + (B * X) + C.

The computer would perform, for a given value of x, the operations indicated below:

$$A * X, (A * X) * X, B * X,$$
$$(A * X * X) + (B * X),$$
$$(A * X * X) + (B * X) + C,$$

in all, five operations – three multiplications and two additions.

There would be no saving if we used the symbol Γ for exponentiation in BASIC, for example,

$$\text{LET } Y = A (X \Gamma 2) + (B* X) + C,$$

since the computer would simply compute 'X Γ 2' by means of the multiplication 'X $*$ X'. If, however, we used the algebraically equivalent command

$$\text{LET } Y = X * ((A * X) + B) + C,$$

the computer would only perform four operations – *two* multiplications and two additions. You can compare these alternatives by choosing values for a, b, c, and x, and executing the operations with a hand calculator.

If the computer does a multiplication in 4×10^{-6} seconds, then the saving achieved by the third command is negligible in computing the polynomial once. But it may be non-trivial if the polynomial must be computed in the order of 10^6 times.

For a polynomial of the third degree,

$$P(x) = ax^3 + bx^2 + cx + d,$$

a step can be saved by computing $x^3 = x \cdot x^2$ from the previously calculated value of x^2. Thus the direct command

$$\text{LET } Y = (A * (X \Gamma 3)) + (B * (X \Gamma 2)) + (C * X) + D$$

means six multiplications and three additions. The alternative

$$\text{LET } Y = X * (X * ((A * X) + B) + C) + D$$

only requires three multiplications and three additions. Now the saving, if the polynomial must be computed many times, is more noticeable.

Exercises

8. Compare alternative programs for computing an arbitrary polynomial of the fourth degree. What is the smallest number of operations needed?

9. Choose values for the coefficients in exercise 8 and compute the polynomial, with a hand calculator, using the various procedures you compared in that exercise. Compare the times for the various alternatives.

10. Find out at any nearby computer facility the computer times for a multiplication and an addition, and the cost of computer time. Decide what amount of money is significant to you. How many computations of a fourth degree polynomial would be needed for your most efficient program to produce a significant saving to you?

11. Work out exercise 8 for fifth degree polynomials. Generalize to polynomials of degree n.

12. Consider a fifth degree polynomial:

$$P(x) = x^5 + Ax^4 + Bx^3 + Cx^2 + Dx + E.$$

(For the sake of simplicity we have taken the coefficient of x^5 to be 1.) Try to find a, b, c, d and e such that, if

$$y = x(x + a) + b,$$

then

$$P(x) = y[x(y + c) + d] + e.$$

If you equate coefficients of like powers of x on both sides, you obtain a system of five equations in five unknowns. It is easy to solve for a. The unknowns c and d can be expressed in terms of b by using the next two equations. If these expressions are used in the next equation, you obtain a quadratic equation for b. The last equation is then easy to solve for e. *After* the values of a, b, c, d, and e are obtained, how many operations are needed to compute $P(X)$? Compare with the best method you found in exercise 4.

This method of computing a polynomial of the fifth degree was discovered by the late T. C. Motzkin. He also found a more efficient method for computing a sixth degree polynomial. The most efficient method of computing an nth degree polynomial is unknown.

13. Choose values for A, B, C, D, and E, and compute $P(x)$ for several values of x by Motzkin's method. How many computations are needed before the *total* amount of work in Motzkin's method is less than the amount required by the method of exercise 11?

In many high school texts a method of 'synthetic division' is given for computing the value of a polynomial. The method is based on the *remainder theorem*:

If the polynomial $P(x)$ is divided by $x - k$, then the remainder is $P(k)$. The proof is very simple. Let the quotient be the polynomial $Q(x)$, and let the remainder be R. We then have the identity

dividend = (quotient · divisor) + remainder,

or

$$P(x) = Q(x) \cdot (x - k) + R.$$

Since the degree of the remainder R must be less than the degree of the divisor $x - k$, which is 1, the remainder must be a constant. If we set $x = k$ in the above identity, then we obtain

$$P(k) = Q(k) \cdot (k - k) + R = R.$$

The method of synthetic division is an efficient way of arranging the process of division. Suppose, for example, that $P(x) = 3x^2 + 4x - 2$, and $k = 5$. We observe that in the division process

$$
\begin{array}{r}
3x + 19 \\
x - 5 \overline{\smash{\big)}\ 3x^2 + 4x - 2} \\
\underline{3x^2 - 15x} \\
19x - 2 \\
\underline{19x - 95} \\
93
\end{array}
$$

the powers of x merely serve as place-holders, and that the essential calculations involve only the coefficients. We could just as well write only the coefficients, being careful to write each in its proper place:

$$
\begin{array}{r}
3 + 19 \\
1 - 5 \overline{\smash{\big)}\ 3 + 4 - 2} \\
\underline{3 - 15} \\
19 - 2 \\
\underline{19 - 95} \\
93
\end{array}
$$

We observe, however, that there are several unnecessary repetitions in the above arrangement; for example, the 3 occurs in three places. Also the 1 in the divisor will always be there. Furthermore the subtractions could be replaced by the additions of the negatives. This leads us to the more economical arrangement

$$
\begin{array}{r|rrr}
+ 5 & 3 & + 4 & - 2 \\
& & + 15 & + 95 \\
\hline
& & + 19 & + 93
\end{array}
$$

where the quotient $= 3x + 19$, and the remainder $= 93$.

Let us consider the more general problem of dividing the polynomial $P(x) = ax^2 + bx + c$ by $x - k$. The above process leads to the calculation

$$
\begin{array}{r|rrr}
k & a & + b & + c \\
& & + ak & + k(ak + b) \\
\hline
& & ak + b & k(ak + b) + c
\end{array}
$$

The important point for us just now is that the process yields

$$P(k) = R = k(ak + b) + c,$$

which is exactly what we obtain by our previous method where we put $P(x)$ in the form

$$P(x) = x(ax + b) + c.$$

Thus we see that for quadratic polynomials the algorithm for efficient programming discussed before coincides exactly with the algorithm of synthetic division.

Exercises

14. Divide $P(x) = 5x^3 - 4x + 1$ by $x - 2$, both by the usual method and by synthetic division. Check that the remainder is $P(2)$.
15. Divide the polynomial $P(x) = ax^3 + bx^2 + cx + d$ by $x - k$, by the method of synthetic division. What formula does this give for the remainder?

Polynomials are the simplest type of functions. As we have just seen, they are easy to compute. Therefore it is often useful to approximate other more complicated functions by polynomials.

1.6   Numerical solution of algebraic equations

In dealing with practical problems, we often need to solve an algebraic equation. For a quadratic equation, such as

$$x^2 - tx - 1 = 0,$$

we have a simple algebraic formula

$$x = \frac{t \pm (t^2 + 4)^{\frac{1}{2}}}{2}$$

for the solution.

For equations of the third and fourth degree, such as

$$x^3 - tx - 1 = 0 \tag{1.1}$$

or

$$x^4 - tx - 1 = 0,$$

there are algebraic formulas for the solutions, but they are so complicated that they are usually of no practical value.

For equations of degree greater than four, such as

$$x^5 - tx - 1 = 0, \tag{1.2}$$

there is, in general, no algebraic formula for the solution. In such cases, we are forced to look for other methods of solving the equations.

If, for instance, we take $t = 1$ in (1.1) and calculate a table of values for the polynomial $P(x) = x^3 - x - 1$, we find that $P(1) = -1, P(2) = 5$. Since P is a continuous function, P has a root v in the interval $1 < x < 2$ (fig. 1.5).

One approach for obtaining a better approximation to v is to approximate the polynomial P in the interval $1 \leqslant x \leqslant 2$ by a linear function. The equation of the line joining $(1, -1)$ to $(2, 5)$ is

$$\frac{y - (-1)}{x - 1} = \frac{5 - (-1)}{2 - 1} \text{ , or } y + 1 = 6(x - 1).$$

If we set $y = 0$ and solve for x, we obtain

$$x = 1 + \tfrac{1}{6} \sim 1.167$$

for the x-coordinate of the intersection of the line with the y-axis. We expect this to be a fairly good approximation to v. The value of P at this point,

$$P(1.167) = -0.579,$$

is a good measure of how good an approximation 1.167 is to v.

We can repeat this process to obtain a better approximation. Since $P(1.167)$ is negative v is between 1.167 and 2, that is, $1.167 < v < 2$. We find the equation of the line joining $(1.167, -0.579)$ to $(2, 5)$,

$$y + 0.579 = \frac{5 - (-0.579)}{2 - 1.167} (x - 1.167),$$

and find its intersection with the x-axis by setting $y = 0$ and solving for x. We obtain

$$x = 1.167 + (0.579) \frac{(2 - 1.167)}{(5 + 0.579)}.$$

Fig. 1.5.

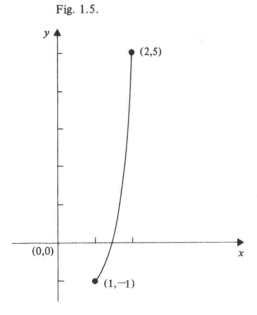

In general, if $P(x_1)$ and $P(x_2)$ have opposite signs and $x_1 < x_2$, then P has a root v such that $x_1 < v < x_2$. The equation of the line joining $(x_1, P(x_1))$ to $(x_2, P(x_2))$ is

$$y - P(x_1) = \left[\frac{P(x_2) - P(x_1)}{x_2 - x_1} \right] (x - x_1).$$

If we set $y = 0$ and solve for x we obtain

$$x = x_1 - P(x_1) \left[\frac{x_2 - x_1}{P(x_2) - P(x_1)} \right] \tag{1.3}$$

as our next approximation. We test this approximation by computing $P(x)$. If $P(x) = 0$, then $x = v$ is the desired root. If $P(x)$ has the same sign as $P(x_1)$, that is, $P(x)P(x_1) > 0$, then we take this as our new x_1. If $P(x)$ and $P(x_1)$ have opposite signs, then we take x as our new x_2. This process is often called the 'method of false position'.

Exercises

16. Calculate the next two approximations to v in the above example, by the method of equation (1.3).
17. Apply the above process to equation (1.2), with $t = 1$, and obtain a good approximation to the positive root v. How do we know that there is a unique positive root?
18. In exercises 16 and 17 do you get good lower bounds for the root v? Do you get good upper bounds? How can you get better upper bounds?
19. Apply the above process to equation (1.2) with $t = 16$ and obtain a good approximation to the unique positive root. Graph the polynomial carefully. How many negative roots are there? Obtain approximations to them.

If we know numbers a and b such that $P(a)P(b) < 0$, then we can construct a flow-chart for the above process. The schematic chart shown in fig. 1.6 is almost good enough. The only trouble is that you will rarely have the good luck of finding x such that $P(x) = 0$ exactly. Thus theoretically the process described by the chart may go on for ever. Also, for practical purposes, you do not need to find v exactly. Therefore you should decide how close you want $P(x)$ to be to zero. If you are satisfied with

$$|P(x)| < 0.001,$$

then you should replace the question 'Is $y = 0$?' by

Is $|y| < 0.001$?

In the above examples, $P(0) = -1 < 0$, and $P(x) > 0$ for large values of x.

Hence we can compute $P(x)$ for $x = 0,1,2,\ldots$, until we come to the first integer at which P is positive. We may take this as b and $b - 1$ as a to get started on the above chart. For more general polynomials there is no good general method for finding suitable values for a and b. The simplest way to start is to make a graph or a table of values for the polynomial P. It is often helpful (for people who know calculus) to examine the derivatives P' and P'', and to find or estimate the points where P has maxima or minima.

Exercises

20. For $t = 3$ in equation (1.1), find suitable values for a and b. Execute the above program with a hand calculator. How many steps did you need?

Fig. 1.6.

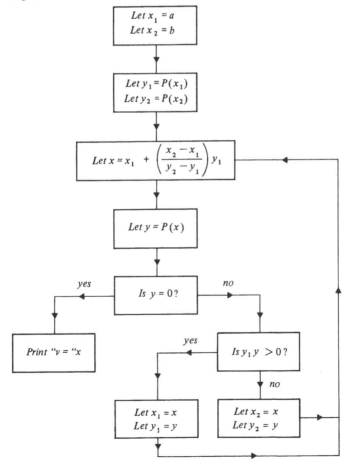

21. Translate fig. 1.6 into a program for solving equation (1.2), given t, a, and b. Start numbering your commands with 100.
22. Write a program for finding suitable a and b for equation (1.2), given t. Combine this with your program for exercise 21.
23. Try to program in exercise 22 for various values of t. What happens if t is large? If t is close to zero?

2

The logarithm

In this chapter we show several ways of introducing logarithmic functions. Napier (1550-1617) invented logarithms originally as an aid to computation, by a method similar to the one we use in section 2.5. Later on, Briggs showed that logarithms to base 10 are more suitable for practical computation. Commercial slide rules exhibit a concrete model of logarithms to base 10. Since the advent of the computer and the hand calculator, logarithms and slide rules are no longer important for purposes of computing, but the logarithmic function is still fundamental for most branches of mathematics and many of its applications.

In sections 2.1-2.3 we give several ways of introducing the logarithmic function in connection with applications. The study of measuring information leads naturally to logarithms to base 2. We focus attention on the basic properties of this function, and especially on how to calculate it using only simple mathematics. This should remove most of the mystery that surrounds the numbers in logarithmic tables in the eyes of most students. The usual texts have an adequate supply of routine problems with which the teacher may supplement the material given here.

Another purpose of sections 2.1-2.3 is to show how a body of knowledge can be organized in the form of a deductive science. We try to give some idea of where postulates come from and how they are related to the real world. The insights of these sections should also be valuable in the teaching of high school geometry.

Sections 2.4 and 2.5 illustrate how the logarithmic function can be introduced at a more basic level. We believe that it is valuable for teachers at more advanced levels to see how simple the fundamental concepts really are.

Section 2.6 gives a treatment suitable for advanced high school or college students. Here we discuss logarithms to an arbitrary base. Until recently, logarithms to base 10 have been the most important for practical computation, while logarithms to base e (natural logarithms) have been, and continue to be, the most important for theoretical purposes and more advanced applications. For work with computers, logarithms to base 2 are nowadays very widely used.

2.1   Coding

The storage and retrieval problem

Suppose you run into a mathematical problem. Should you try to solve it yourself, or should you try to find out whether it has been solved, or at least what work has been done on it? Suppose you want to send in a proposal for a research project to a government agency like the National Science Foundation. Obviously they will not give you a contract if the answer to that problem is known. How will you go about finding the information?

You might first dig into the journal *Mathematical Reviews*, published by the American Mathematical Society, which gives short reviews of all mathematical articles and books published anywhere in the world. You might start with the current issue or volume and work back. The index is of some help if you know the author or the general topic. Under a general topic like irrational numbers, you might find 20 or 30 papers per year listed, but the index would not be so detailed as to have a listing under 'the decimal expression of π'.

During the past few years, *Mathematical Reviews* has published about 15 000 reviews per year. There is, of course, a time lag between publication of results and publication of the review. You can order a microfilm or photocopy of any paper from *Mathematical Reviews* if you know what to ask for.

Imagine, then, that you have a library or an archive with copies of all these papers stored away, and you want to locate all those, if any, which have the information you want. How should the paper be stored or classified so that you can get the information you want out again at reasonable cost of time and money? There have been several international conferences on this problem. A number of libraries have large research grants for applying mathematics and computers to questions like this.

The same problem arises in other fields. Suppose you find an insect and you wish to know whether it belongs to a new species or a known species; if the latter, you will want to identify your insect. How can you search through the descriptions of the million or so known species to get an answer to your question? Again, suppose you are managing a technical aid project, and you ask the National Roster of Scientific and Technical Personnel to locate a civil engineer who knows colloquial Tibetan. How can they store their information about people so as to answer such requests efficiently?

Classification

Let us study the problem of cataloging the mathematical papers as they are published, in such a way that you can recognize the papers you are interested in from the cataloging card. One approach is to ask a standard set of questions in a standard order and record the answers on the card.

Your first question might be 'Is it about algebra, geometry, or probability?' You might then print A, G, or P at a certain spot on the card. Then you might ask 'What number system is used – integers, rationals, reals, or complex numbers?', and so on. Then, when you want to look something up, you, a clerk, or a computer could search through the cards for the kind of papers you want.

For the sake of simplicity, we could record each answer with a single symbol at a certain standard spot on the card. However the first question above would then not be suitable. How would a paper on the solutions of the equation $x^2 + y^2 = z^2$, which also discussed the Pythagorean theorem be classified? A and G cannot both be printed on the same spot.

So the questions should have *mutually exclusive* answers. We should only ask questions which have unique answers. If no question has more than 26 possible answers, we could symbolize the alternatives by letters of the alphabet, as in the familiar multiple-choice tests. If there are never more than ten possible answers, we could use digits $0,1,2,\ldots,9$. If we use only yes–no questions, we could simply punch a hole for 'yes' and not punch for 'no'.

The symbols on the card then contain a certain amount of information: to which one of a certain number of categories does the paper belong? The more possible categories there are, the more information the card contains.

If the answer to the first question is A, B, or C, and to the second is A or B, then a card will have one of the following 'words' printed on it: AA, AB, BA, BB, CA, or CB. The number of possible categories is then the same as the number of different words – in this case six.

Exercises
1. Given three questions with four answers (A, B, C, D) to the first, three (A, B, C) to the second, and two (A, B) to the third, list all the code words. How many categories are there?
2. Given ten yes–no questions, how many categories of code word are there?
3. There are three 10-choice questions. How many categories are there?
4. Which gives more information, three 10-choice questions or ten yes–no questions?

5. Which gives more information, five 3-choice questions or eight 2-choice questions?

6. How many 2-choice questions do you need to identify 1000 categories?

Combinatorics

Suppose we use only 3-choice questions, so that we can code each answer with A, B, or C. Each card will have a code word like $ABBAC$, giving us a classification of the cards. If there are n questions, then how many categories are there? This is equivalent to the problem 'How many sequences of n letters are there, if there are three possible choices for each letter?' Let a_n be the number of such sequences. Let A_1 be the set of all sequences which begin with A, B_1 the set of all which begin with B, and C_1 the set of all which begin with C. Then

$$a_n = N(A_1) + N(B_1) + N(C_1),$$

where $N(A_1)$ is the number of members in the set A_1. Any sequence in A_1 is obtained by writing A, then a sequence of $n-1$ letters, with three possible choices for each letter. Therefore we have

$$N(A_1) = a_{n-1},$$

and similarly

$$N(B_1) = N(C_1) = a_{n-1}.$$

It follows that

$$a_n = 3a_{n-1}$$

and

$$a_1 = 3.$$

We can now compute a_n successively for $n = 2, 3, \ldots$:

$$a_2 = 3a_1 = 3 \times 3 = 3^2 = 9,$$
$$a_3 = 3a_2 = 3 \times 3^2 = 3^3 = 27,$$

and in general

$$a_n = 3^n.$$

Exercises

7. With an 'alphabet' of two letters, say the digits 0 and 1, how many n-letter 'words' are there? List all of them (0000, 0001, etc.) for $n = 4$.

8. With a 10-letter alphabet, how many n-letter words are there?

9. The outcome of n rolls of a die can be described by an n-letter word,

using the 6-letter alphabet {1,2,3,4,5,6}. How many possible outcomes are there?

10. How many 7-digit telephone numbers are possible? How many digits would be needed to have a number for every person in the United States?

11. Which gives more information, three 3-choice questions and four 5-choice questions, or seven 4-choice questions?

2.2   Measuring information

Amounts of information

In 1948 the mathematical engineer Claude Shannon, then with Bell Laboratories, invented a new branch of mathematics called information theory. This theory has been applied since then to electrical engineering, psychology, linguistics, and even to library work.

Shannon was interested in comparing various methods for coding, transmitting, receiving, and decoding information. In order to find out which methods are most efficient, he needed a way to measure amounts of information. We shall study here the simplest case of this problem of measuring information.

We take as the unit of information the bit, which is the amount of information in the answer to a 2-choice question. The question may be whether a certain digit is 0 or 1, whether a light bulb is on or off, whether a hole is or is not punched at a certain spot, etc.

Obviously the amount of information obtained from the answer to a 3-choice question is *more* than the amount of information in the answer to a 2-choice question, as more possibilities are eliminated. We shall call this amount $I(3)$, so

$$I(3) > I(2)$$

or, since

$$I(2) = 1,$$
$$I(3) > 1.$$

Let $I(n)$ be the amount of information we have if we know the correct answer out of n possibilities. Then, if $m < n$, the same reasoning as above shows that

$$I(m) < I(n).$$

Can one add amounts of information? Suppose that we have answers to a 3-choice question and a 5-choice question. How much do we know altogether? Since there are three possible answers to the first question and five to the second, there are 3 × 5 possible cases altogether. So the answers to the two questions give us $I(15)$ bits of information, or

$I(3) + I(5) = I(15).$

This reasoning shows in general that

$I(m) + I(n) = I(mn).$

Let us group our main results together:

$I(2) = 1.$ (2.1)

If $m < n, I(m) < I(n).$ (2.2)

$I(mn) = I(m) + I(n).$ (2.3)

In arriving at these conditions, we have assumed that there is a measure of amounts of information, and we have simplified the problem by ignoring the content of questions. In (2.3) we tacitly assumed that the questions under consideration are *independent*, that is, the answer to one question has no effect on the answer to the other. In defining $I(n)$, we have also tacitly assumed that all the possible answers to the n-choice question are equally likely. In any attempt to describe mathematically a situation in the real world, we have to make simplifications and idealizations like these in order to get started. After developing a simple theory on the basis of such simplifications, we can then try to modify it to obtain a more sophisticated, but more realistic theory. This will usually be at the cost of making the theory more complicated.

Calculating $I(n)$

Principles (2.1), (2.2), and (2.3) enable you to calculate $I(n)$ as accurately as you please. For example, let us estimate $I(5)$. Since

$2^2 < 5 < 2^3$

we have, by (2.2),

$I(2^2) < I(5) < I(2^3),$

but by (2.3) we see that

$I(2^2) = I(2 \times 2) = I(2) + I(2) = (\quad)$

by (2.1), and

$I(2^3) = I(2 \times 2 \times 2) = (\quad).$

Fill in the missing values in these expressions and in the following text. Therefore we find that

$2 < I(5) < 3,$

that is, the answer to a 5-choice question contains between 2 and 3 bits of information.

We can obtain a more accurate estimate if we compare 5^2 with the successive powers of 2:

$$2^4 = (\quad), 2^5 = (\quad), 2^6 = (\quad), 2^7 = (\quad), \text{etc.}$$

We find that

$$(\quad) < I(5^2) = 2I(5) < (\quad),$$

so that

$$(\quad) < I(5) < (\quad).$$

One side of this inequality improves our previous estimate.

We obtain a better squeeze on $I(5)$ by comparing 5^3 with the successive powers of 2:

$$2^? < 5^3 < 2^?,$$

which yields

$$(\quad) < 3I(5) < (\quad),$$

and

$$(\quad) < I(5) < (\quad).$$

We could proceed in this way. If we find k such that

$$2^k < 5^{1000} < 2^{k+1}$$

then

$$k < 1000I(5) < k + 1,$$

and

$$k/1000 < I(5) < k/1000 + 1/1000,$$

so that $I(5)$ is determined to within an error of less than 0.001.

Exercises

12. Make a table of powers of 2 up to 2^{20}. Calculate 3^{10} and obtain an estimate of $I(3)$.
13. Calculate 7^6 and estimate $I(7)$.
14. Find n such that

$$\left(\frac{2^7}{11^2}\right)^n < \frac{11}{2^3} = 1.375 < \left(\frac{2^7}{11^2}\right)^{n+1}$$

and estimate $I(11)$.
15. (*a*) Which is larger,

$$\left(1 + \frac{1}{100}\right)^n \quad \text{or} \quad \left(1 + \frac{n}{100}\right)?$$

(*b*) Is there an n such that $(1.01)^n > 2$?

More computations

By now it is quite obvious that

$$I(n) = \log_2(n),$$

and the next stage is to look for better and quicker ways to compute $I(n)$. As an example, let us look at $I(5)$, and set $I(5) = x$.

First let us estimate x between consecutive integers. In our previous work, we found that $2 < x < 3$. Then we set

$$x = 2 + \frac{1}{x_1}$$

where $x_1 > 1$. We obtain

$$I(5) - 2 = \frac{1}{x_1},$$

or

$$x_1[I(5) - I(4)] = 1.$$

We try to estimate x_1 between consecutive integers. Let us compare $I(5) - I(4)$ with $I(2) = 1$. We see that

$$2[I(5) - I(4)] = 2I(5) - 2I(4) = I(5^2) - I(4^2),$$

which is less than $I(2)$ if

$$I(5^2) < [I(4^2) + I(2)] = I(4^2 \times 2).$$

Is this so? What about

and

$$3[I(5) - I(4)] = 3I(5) - 3I(4) = I(5^3) - I(4^3),$$
$$4[I(5) - I(4)] = I(5^4) - I(4^4)?$$

We find that

$$3[I(5) - I(4)] < 1 < 4[I(5) - I(4)],$$

and conclude that

$$3 < x_1 < 4.$$

Hence

$$2.25 = \left(2 + \frac{1}{4}\right) < x < \left(2 + \frac{1}{3}\right) = 2.33,$$

so that

$$I(5) = 2.29 \pm 0.04.$$

If we want to refine this estimate, we put

$$x_1 = 3 + \frac{1}{x_2}$$

where $x_2 > 1$, and obtain the equation

$$x_2[I(2^7) - I(5^3)] = I(5) - I(2^2). \tag{2.4}$$

(Hint; by substituting $(3 + 1/x_2)$ for x_1 in the equation

$$x_1 [I(5) - I(4)] = 1$$

we get

$$3x_2 [I(5) - I(4)] + [I(5) - I(4)] = x_2.)$$

Now we simply continue in the same way.

More properties of $I(n)$

We can simplify the calculations by defining

$$I\left(\frac{m}{n}\right) = I(m) - I(n);$$

for example,

$$I\left(\frac{128}{125}\right) = I(128) - I(125).$$

Let us first check that this definition is legitimate. A rational number has many representations as a fraction:

$$\frac{5}{4} = \frac{10}{8} = \frac{15}{12} = \frac{500}{400}, \text{etc.}$$

This yields the following 'values' for $I(\frac{5}{4})$:

$$I(5) - I(4), I(10) - I(8), I(15) - I(12), \text{etc.}$$

Are all these numbers really equal? Is, for instance, $I(5) - I(4)$ equal to $I(15) - I(12)$? Compare $I(5) + I(12)$ with $I(4) + I(15)$, using (2.3).

Exercises

16. If a, b, c and d are positive integers and $a/b = c/d$, compute

    $$[I(a) - I(b)] - [I(c) - I(d)] .$$

    We define

    $$I\left(\frac{a}{b}\right) = I(a) - I(b).$$

    Does this definition assign a unique value to $I(r)$ when r is rational?
17. Suppose that a, b, c and d are positive integers and $a/b < c/d$.
    (a) What is the sign of $bc - ad$?
    (b) Compute

    $$I\left(\frac{c}{d}\right) - I\left(\frac{a}{b}\right)$$

    using the definition in exercise 16. What is the sign of this difference?
    (c) Is principle (2.2) valid for rational numbers?

18. Suppose a, b, c, and d are positive integers, and

$$r = \frac{a}{b}, s = \frac{c}{d}.$$

  (a) Compute $I(r) + I(s)$ and $I(rs)$.

  (b) Does principle (2.3) hold for rational numbers?

19. (a) Compare the successive powers of

$$\frac{128}{125} = 1.024 \quad \text{with} \quad \frac{5}{4} = 1.25,$$

  that is, find an integer n such that

$$(1.024)^n < 1.25 < (1.024)^{n+1},$$

  and estimate x_2 in equation (2.4).

  (b) Use this result to estimate x_1, then x. What is the difference between the upper and lower estimates you obtain?

20. Use the method from the text to estimate $I(3), I(7)$, and $I(11)$.

21. What integers from 2 to 100 have only 2, 3, 5, 7 or 11 as factors? Use this information to make a rough table of $I(n)$ for $2 \leqslant n \leqslant 100$.

22. Does principle (2.3) work if $n = 1$? What must $I(1)$ be? What is the interpretation of this result?

This section illustrates a typical phenomenon in applied mathematics. Often a mathematical model for a situation in the real world takes on a 'life' of its own. The analysis of the model often leads us to problems and concepts which have no obvious concrete interpretation. The introduction of $I(r)$ for rational values of r is an example of this. On the other hand, sometimes concepts which arise in the mathematical analysis suggest new and useful concepts in the real world situation. Thus the physical concepts of energy and temperature originated in the analysis of certain mathematical models for physical phenomena.

2.3 Comments on models and axioms

In this chapter we have been examining the problem of measuring information. We started by assuming that there is such a measure, represented by the function $I(n)$. By analyzing the properties which we think the measure ought to have, we were led to conditions (2.1), (2.2), and (2.3). If we take these as *postulates* and take the function I to be undefined, we obtain an abstract mathematical system, which we call a *mathematical model*, for information theory. This set of postulates, together with their logical consequences, is called a *deductive science*. Our intended interpretation is in terms of information, but there may be other interpretations as well.

In general, when we are faced with a phenomenon in the real world, we first try to investigate it by observation and experimentation. After we have built up a body of knowledge, we try to organize it in such a way that we can make predictions. Usually this requires that we set up a mathematical model to describe the situation under consideration. The assumptions and the basic concepts (the undefined terms) of the mathematical model arise from the intended concrete interpretation. The theorems will usually be predictions of the results of further experiments.

The real world is too complicated for our finite human minds to grasp fully, so the construction of the mathematical model involves simplifications and idealizations. Thus, in the above model for measuring information, we assumed that all the answers to a question are equally likely and gave them equal weight. A more sophisticated, but more complicated, model would allow the questions to be assigned different weights. For a development of such a theory see A. Y. Khinchin, *Mathematical foundations of information theory* (Dover Publications, New York, 1957). The important point to notice here is that the choice of the postulates and undefined terms is a human decision.

Most mathematical systems of interest arose from such attempts to describe aspects of the real world. Euclidean geometry is a classical example of this type of applied mathematics.

The historical development of geometry suggests also the best approach to teaching it. Just as centuries of empirical work in Egypt and Babylonia preceded the emergence of the deductive approach in the century before Euclid, so a good deal of observation and measurement of geometric figures should come before the introduction of postulates. The postulates should arise as generalizations of the students' experience.

Similarly, the basic concepts such as 'measure of information', 'point', and 'line' are simplifications and idealizations. One should emphasize that one does not see actual 'points' and 'lines' in the real world. The Danish mathematician Hjelmslev published a system of postulates describing 'points' as dots of finite size and 'lines' as streaks of finite width, restricted to a finite piece of paper (J. T. Hjelmslev, 'La géométrie sensible II', *Enseignement Mathématique*, 38 (1942), 294-322). This geometry is more realistic than Euclidean geometry but is much more complicated. The students should be aware that the assumptions and basic concepts are matters of human choice, and can be changed.

The theorems can be considered as predictions. When we test such predictions by further observation, we find out to what extent our model actually fits the real world it is meant to describe. Thus the theorem that the medians of a triangle are concurrent is a striking prediction which should be tested by actual construction. The Pythagorean theorem should be tested by measurement.

Unfortunately most school textbooks present the postulates of geometry as though they are revealed from on high, without any discussion of where they came from, or of any alternatives. Also, in most school curricula, students in elementary school and junior high school get inadequate experience with the empirical facts of geometry. Hence they are not sufficiently prepared for the formal organization of geometry as a deductive science in high school. Therefore it is left to teachers to fill in these gaps as well as they can.

It would be a good exercise for the students to try to formulate a mathematical model for Hjelmslev's geometry of dots and streaks.

2.4   The game of 'Info'

Starting the game
The idea of measuring information, and the function $I(n)$, can be introduced at a more basic level via a variant of the familiar game of 'Twenty Questions'. We call our game 'Info'.

A set of objects, such as the students in the class or the integers from 1 to 100 inclusive, is agreed upon in advance. Two players play. One player thinks of an object in the set, writes its name on a slip of paper, and hands it to an umpire.

The other player must guess the object the first player is thinking of. He may ask only questions which can be answered by 'yes' or 'no'.

When the second player succeeds, the players interchange roles. The second player chooses a member of the set and the first player must guess it. Whoever asks the smallest number of questions wins.

In teaching 'Info', we have found it useful, after some free play, to play against the class. We think of a number from 1 to 100 and write it on a slip of paper. As the pupils ask their questions, we write them on the blackboard together with the answers. When the class has guessed our number, we step out of the room while the class agrees on a number. They send a messenger to call us back, and we ask our questions. As before, we write the questions and answers on the blackboard.

There are several methods we can use. The one which can be used to teach the most is to ask 'Is the number odd?' If the answer is 'No', we say 'Divide the number by 2. Is the result odd?' If the answer is 'yes', we say 'Subtract 1, then divide by 2. Is that result odd?' We repeat the procedure until we have asked seven questions. Then we name the class's number.

After we beat the class twice by such a strategy, it becomes obvious that we have a method which is better than random guessing. The class is then receptive to a discussion of why some questions are better than others. For example, it almost always happens that some of their questions are super-fluous. Also, the question 'Is the number even or odd?' always has the answer 'yes', and therefore gives no information! The question 'Is the number between 10 and 20?' is somewhat ambiguous because it is not quite clear whether 10 and 20 are included or not. One should add either 'inclusively' or 'exclusively'. These remarks already teach the importance of *saying what you mean.*

The fact that one can *predict* the answers to some questions from those of the preceding questions teaches that questions have logical relations. For example, the answer 'yes' to 'Is the number divisible by 10?' tells you what the answer to 'Is the number odd?' must be.

One way of showing this clearly is to list, after each of the class's questions the set of possibilities that remain. Thus, we might have:

(*a*) Is the number greater than 70? No $\{1,2,\ldots,70\}$.

(*b*) Is the number divisible by 5? Yes $\{5,10,15,20,25,30,35,40,45,50,55, 60,65,70\}$.

(*c*) Is the number even? No $\{5,15,25,35,45,55,65\}$.

(*d*) Is the number greater than 50? No $\{5,15,25,35,45\}$, etc.

This procedure now makes it clear when the answer to a question can be determined from the preceding answers.

At this point it is good teaching strategy to give the pupils, for homework, the assignment of playing 'Info' with friends and family, and trying to discover further principles for asking good questions.

Exercises

23. Which is a better first question, 'Is the number 17?' or 'Is the number greater than 70?' Why?

24. Suppose the answer to question (*b*) in the textual example above had been 'no'. List the set of possibilities then remaining after each question.

25. Is question (*b*) a good question? Which of the following would be a better question:

    (*b′*) Is the number greater than 35?

    (*b″*) Is the number divisible by 10?

    (*b‴*) Does the number have two digits?

26. Can any of the other questions in the example be improved? How?

27. How does the method we used work? Try the method on the numbers from 1 to 8 inclusive. Consider whether 0 is even or odd. What can our method teach, in addition to logarithms?

28. Can you devise a simpler method than ours which will also give the desired number after seven questions?

Analyzing the game

In the second lesson, after the pupils have reported on the discoveries they have made, you can begin an analysis of what a question does by continuing to study the above example. Question (*a*) separates the set of all possibilities $\{1,2,\ldots,100\}$ into two *subsets*, a yes-set $\{71,72,\ldots,99,100\}$ and a no-set $\{1,2,3,\ldots,70\}$. These sets for question (*b*) are:

yes-set $\{5,10,15,20,\ldots,100\}$,
no-set $\{1,2,3,4,6,7,8,9,\ldots,94,96,97,98,99\}$.

The set of all possibilities remaining after the first two questions is what is common to the no-set of question (*a*) and the yes-set of question (*b*), that is, their *intersection*.

Let us return to question (*a*). If the answer is 'yes', *how many* possibilities remain? What about if the answer is 'no'? What is the *worst* that can happen? What is the *maximum risk* in asking question (*a*)?

Would it be better to ask question (*b*) first? Now what is the worst that can happen? What is the maximum risk in asking question (*b*)? As a first question, which is better, (*a*) or (*b*)? Can you think of a better question than either? Can you think of a question which *minimizes* the maximum risk? Can you think of another? Can such a question be improved?

Now can you see why our strategy worked? Return to exercise 28 and try it again.

Exercises

29. Make a table of *n*, the number of possibilities in the original set, versus *Q*, the number of questions needed in the best strategy:

n	50	100	200	300	400	500	512	513	1000
Q		7							

30. For each number *Q* of questions, what is the *largest n* for which *Q* questions are enough? Make a table:

Q	1	2	3	4	5	6	7	8	9	10
n										

We can express the above results in another, more suggestive form. In the study of information, we take as our unit for measuring amounts of information the *bit*, which is the amount of information in the answer to a 2-choice (true-or-false) question. We may interpret the result in the second column of the table in exercise 29 thus:

To find the correct answer among 100 possibilities,
we need at most 7 bits of information.

We may think of finding the correct number in the set $\{1, 2, \ldots, 100\}$ as finding the answer to the 100-choice question

Is it 1 or 2 or 3 or ... or 100?

So we can also express the result in the form

The answer to a 100-choice question contains at most 7 bits of information.

The seventh column in the table of exercise 30 gives the more precise result

The answer to a 128-choice question contains exactly 7 bits of information.

Multiple-choice questions

Up to now we have worked with yes-or-no questions, that is, *2-choice* questions. Children will be familiar with *multiple-choice* questions from their examinations. Typical 3-choice questions might be:

(*a*) The number is (i) less than 10; (ii) at least 10 but less than 50; (iii) at least 50.

(*b*) The number, on division by 5, gives a remainder of (i) 0 or 1; (ii) 2 or 3; (iii) 4.

Again, each question divides the set of all possibilities into subsets. Suppose again that the set of all possibilities consists of the integers from 1 to 100; list the subsets and find the maximum risk for each of these questions (*a*) and (*b*). Which is a better question? For which kind of 3-choice question is the maximum risk a minimum? Give an example of an optimal 3-choice question.

Make a table of n, the number of possibilities, and Q, the number of 3-choice questions needed in the best strategy:

n	50	200	300	400	500	512	729	730	1000
Q									

Make a table of the number Q of 3-choice questions versus the largest number n of possibilities for which Q questions are enough.

Combine these results with those of exercise 29 into one table:

n	Number of questions needed	
	2-choice	3-choice
. .	.	
. .	.	

Exercises

31. From the above table you can make some comparisons: five 3-choice questions give more information than seven 2-choice questions, but less than eight 2-choice questions. Make similar comparisons for seven 3-choice questions and twelve 3-choice questions.

32. Complete table 2.1. Where have you seen a table like this before?

Table 2.1

Number of 3-choice questions	2-choice questions	
	More than	Less than
1		
2		
3		
4		
5	7	8
6		
7		
8		
9		
10		

33. Make a similar comparison between 5-choice and 2-choice questions.

We can express the result of exercise 31 in the form

The answers to five 3-choice questions give between 7 and 8 bits of information

or

The answer to a 3-choice question contains between $\frac{7}{5} = 1.4$ and $\frac{8}{5} = 1.6$ bits of information.

Thus we have introduced, in a very basic way, the concept of $\log_2(3)$ and have obtained an approximation to its value. In section 2.6 we shall discuss this idea in detail at a more advanced level.

2.5   A class project: the slide rule

Until the development of cheap digital hand calculators, most engineers and many scientists used a calculating device called a *slide rule*. Indeed, one could usually recognize an engineer by the slide rule he carried.

Nowadays the slide rule is obsolete. In fact, the leading manufacturer of slide rules went out of business a few years ago. Nevertheless, the slide rule has several advantages for teaching purposes. For the student the hand calculator is a *black box* – its mechanism is hidden and there is no way for the user

to see how the results are obtained. On the other hand, the slide rule is open
to inspection and, with some study, the user can find out exactly how it
works. It can also be used to exhibit a concrete representation of the logarithmic function.

It is a useful and interesting project for a class or a mathematics club to
make their own slide rule. Afterwards, the students may compare their home-
made slide rule with a commercial one, if one is still available. We have even
taught this material to children in grades 5–6, as a vehicle for giving motivated
practice in computation with fractions and inequalities. For older students, it
can be an interesting challenge to try to find explanations for the fascinating
properties of the fractions which arise in the work described below. We shall
not, however, enter into this aspect, but shall focus attention on what is relevant to the study of the logarithmic function.

Scales

We first remind our pupils that a scale is really a one-to-one correspondence between *points* and *numbers*. This can be done at a basic level by
passing out rulers to the pupils and asking them to observe that the edge of
the ruler bears a scale.

Then we notice that

■   Principle A Equal differences correspond to equal distances.

For example, two points 3 cm apart, such as 1 and 4 or 5 and 8, always correspond to numbers which differ by 3:

$$4 - 1 = 8 - 5 = 3$$

This makes it possible to use two rulers to add or subtract numbers. For
example, fig. 2.1 shows how to add or subtract 5. Work out the rules!

Fig. 2.1

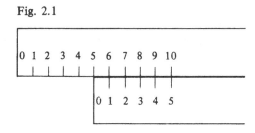

We now propose the following problem: can we make a scale, a one-to-one
correspondence between points and numbers, which satisfies principle B?

■   Principle B Equal distances correspond to equal ratios.

For example, we might want numbers whose ratio is 2,

$$2/1 = 4/2 = 6/3 = 8/4 = 2,$$

to be always 1 cm apart.

Fig. 2.2

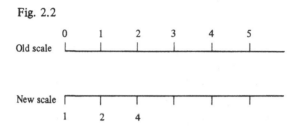

In fig. 2.2 we have begun to label the points on the new scale. Which numbers correspond to the third and fourth points of the old scale? Ask the students to label the other points opposite the integers on the old scale.

You will, of course, recognize the numbers on the new scale as the powers of 2. Just as we could use the old scale for addition, so the new scale can be used to multiply and divide. Note that we are using the laws of exponents here, but the pupils do not need to know them as such at this point. Fig. 2.3 shows us, for instance, how to multiply or divide by 4. When do you read the

Fig. 2.3

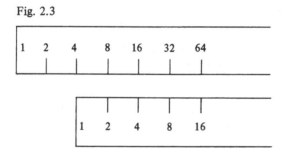

result upwards? When do you read it downwards? Work out the rules!

Since it is agreed that multiplication and division are 'harder' than addition and subtraction, we can all appreciate the value of the new scale. The drawback, however, is that we have only located some very special numbers on the new scale. Which point, for example, should be labeled 3?

The 3-point on the new scale

Let us look again at fig. 2.2. The 3-point on the new scale must be opposite some point between 1 and 2 on the old scale. Let us call its unknown value on the old scale x, as shown in fig. 2.4. The points 1 and 3 on the new scale have the ratio $3/1 = 3$, which corresponds to a distance of x cm. By

Fig. 2.4

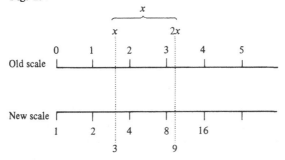

principle B, a distance of x cm must always correspond to the same ratio of 3 on the new scale.

Therefore, the $2x$-point on the old scale, which we get by measuring another 3 cm along the scale, must correspond to a number whose ratio to 3 is 3. Obviously this number is 9, since $9/3 = 3$. As 9 is between 8 and 16, so $2x$ must lie between 3 and 4.

Measuring off another x cm gives us the point $3 \times 9 = 27$ on the new scale, corresponding to $3x$. Since 27 is between 16 and 32, $3x$ is between 4 and 5. Continuing in this way, we obtain the results that are summarized in table 2.2.

Table 2.2

Name of unknown point on old scale	Corresponding value on new scale	Limiting values for point	
		on new scale	on old scale
x	3	2 and 4	1 and 2
$2x$	9	8 and 16	3 and 4
$3x$	27	16 and 32	4 and 5
$4x$	81	64 and 128	6 and 7
$5x$	243	128 and 256	7 and 8
$6x$	729	512 and 1024	9 and 10
$7x$	2187	2058 and 4116	11 and 12

Inequalities

In table 2.2 the values of nx have been sandwiched between other numbers. Let us look at these 'sandwiches' more carefully.

If $2x$ is greater than 3, then x must be greater than $3/2$. Similarly, if $2x$ is less than 4, then x must be less than 2. So we have

$$3/2 < x < 2.$$

(This is maybe the right moment to introduce the notation '<', if the pupils do not yet know it.) Looking at the other rows we then get, in succession,

$$1 < x < 2$$
$$3/2 < x < 2$$
$$4/3 < x < 5/3$$
$$3/2 < x < 7/4.$$

We notice immediately that some of the inequalities do not tell us anything new. For example, the first line tells us that x is less than 2, so the right-hand side of the second line does not add any useful information.

It will help us understand better what is going on if we look at the limiting values obtained for x as points between 1 and 2 on the old scale, as in fig. 2.5.

Fig. 2.5

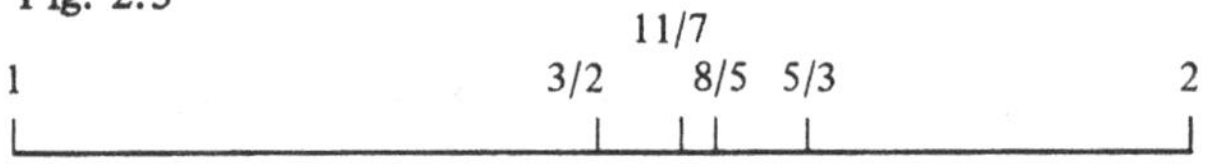

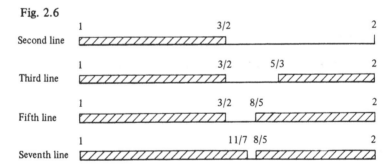

The first line of table 2.2 tells us that x lies somewhere in this interval. The second line restricts the possible interval for x to between 3/2 and 2. In fig. 2.5 the 3/2-point is marked off on the segment between 1 and 2. In fig. 2.6 we have shaded the part of the segment between 1 and 2 that is not relevant any more. This figure shows that our knowledge about x has improved, since

Fig. 2.6

now the length of the segment in which x can lie is restricted to

$$2 - 3/2 = 1/2.$$

The third line of the table restricts us to the interval 4/3 to 5/3; 4/3 does not add to our knowledge, since $4/3 < 3/2$. (This might be the place to remind the students of the use of the common denominator, for $4/3 = 8/6$, whereas $3/2 = 9/6$.) So we obtain $3/2 < x < 5/3$. This interval is of length

$$5/3 - 3/2 = 1/6.$$

You can work with the other lines in the same way. We have marked off the relevant points of fig. 2.5, and shaded the discarded intervals in fig. 2.6.

Table 2.3

Interval	Length
$1 < x < 2$	1
$3/2 < x < 2$	1/2
$3/2 < x < 5/3$	1/6
$3/2 < x < 8/5$	
$11/7 < x < 8/5$	

Work out for yourself the lengths of the successive intervals in table 2.3. Do you notice anything *special* about the fractions in the length column?

Compare each new fraction in fig. 2.5 with its two neighbors:

3/2 with 1 and 2
5/3 with 3/2 and 2
8/5 with 3/2 and 5/3
11/7 with 3/2 and 8/5.

Do you notice any relation between these fractions? It is more obvious in the last two lines. Does it work also for the first two lines? (Write 1/1 for 1, 2/1 for 2.) Can you *predict* what the next new fraction will be? Test your predictions by working out further lines in the appropriate tables.

Exercises

34. Let y on the old scale correspond to 5 on the new scale. Work out estimates for y, as we did for x.

35. Let z on the old scale correspond to 7 on the new scale. Work out estimates for z.

   In teaching, assign each of these two exercises to half the class.

36. Is there anything special about the lengths of the intervals which arise in exercises 34 and 35? What about the relation of each new fraction to its two neighbors? Can this be used to shorten the work?

The slide rule

We now have located several points on the new scale (see fig. 2.7). Do we really need to calculate where 6 goes on the new scale? We notice that $6/3 = 2$, and that the ratio 2 corresponds to 1 cm. Thus the 6-point on the new scale is 1 cm from the 3-point, or

   6 corresponds to $x + 1$.

In the same way the 10-point is (   ) from the 5-point, so that 10 corresponds to (   ). Fill in the missing values. Thus the next point we have to

Fig. 2.7

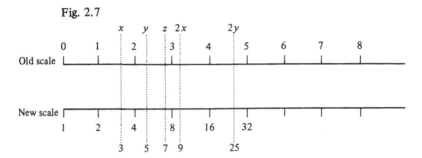

work out by the long method is the 11-point. What about the 12- and 13-points? A careful study shows that we only need to locate the points marked

$$13, 17, 19, 23, 29, \ldots$$

on the new scale, and we can figure out the location of all the rest from these. What is special about the numbers

$$3, 5, 7, 11, 13, 17, 19, 23, 29, \ldots?$$

How are they different from

$$9, 15, 21, 25, 27, \ldots?$$

At this stage we see that it is only necessary to locate prime-numbered points on the new scale. We can now parcel out the next few primes to groups of pupils, say three to five pupils for each prime. At the next meeting of the class we call for group reports and combine the results into a single scale, producing in this way a home-made slide rule. Another use of scales is given in section 2.6.

2.6 The functional equation

There are in fact three main ways to introduce the notion of the logarithm:

(*a*) As an inverse to the exponential function: if $y = a^x$, then $x = \log_a y$.

(*b*) As a definite integral $\log_e x = \displaystyle\int_1^x \frac{dt}{t}$.

(*c*) As a correspondence between additions and multiplications.

The last of these three is historically the first, but it has lately fallen into disfavor and is not used any more at the college level. We will develop the functional equation of the logarithmic function from this point of view, and show how to compute values of logarithms in this manner without using calculus.

The functional equation

Our starting point is the problem of the two scales in section 2.5. For any number x $(x \geqslant 1)$, let $L(x)$ be the number on the top scale in fig. 2.8 which corresponds to x on the bottom scale. Our basic hypothesis is that

On the L-scale equal distances correspond to equal ratios.

Fig. 2.8

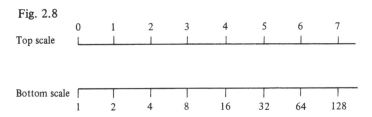

We can express this symbolically (see also fig. 2.9): let a, b, c, d all be $\geqslant 1$ and

$$\frac{b}{a} = \frac{d}{c} \tag{2.5}$$

Fig. 2.9

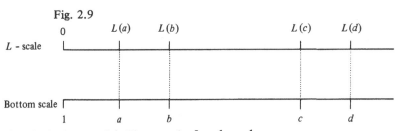

(on the bottom scale). Then on the L-scale we have

$$L(b) - L(a) = L(d) - L(c) \tag{2.6}$$

If we take $d = x$ and $c = 1$, we have $b = ax$, and equation (2.6) becomes

$$L(ax) - L(a) = L(x) - L(1) = L(x) - 0 = L(x),$$

or

$$L(ax) = L(a) + L(x). \tag{2.7}$$

We should also express the fact that order on the bottom scale agrees with order on the top scale:

If $x < y$, then $L(x) < L(y)$. $\tag{2.8}$

Equation (2.7) and inequality (2.8) may be compared with the postulates in the model for information theory (see p. 30).

We need, finally, to express our choice of the ratio which corresponds to a unit distance:

$$L(2) = 1. \tag{2.9}$$

We can now derive estimates as before. If $L(3) = x$, then

$$L(3^2) = L(3 \times 3) = L(3) + L(3) = 2x,$$

and so on. Therefore $L(3^n) = nx$.

Since $L(2^n) = n$, we can now obtain estimates for x. For example, if we locate 3^{100} between two consecutive powers of 2, say

$$2^m < 3^{100} < 2^{m+1},$$

then we apply (2.8) and obtain

$$m < 100\,x < m + 1,$$

so that

$$\frac{m}{100} < x < \frac{m+1}{100}.$$

This determines x, and $L(3)$, to within an error of 0.01.

To determine x to within as small an error as we choose, we have to use larger powers of 3. Therefore we can conclude that conditions (2.7), (2.8) and (2.9) *determine the value of $L(3)$ exactly*. In other words, if both L_1 and L_2 are functions satisfying (2.7)–(2.9), then

$$L_1(3) = L_2(3).$$

The same reasoning can be applied to any other number instead of 3. So we arrive at the conclusion that *there is at most one function L satisfying conditions (2.7), (2.8), and (2.9).*

At this point pause for a moment and consider whether $L(1)$ can be uniquely defined, and what value should be assigned to it.

Exercises

37. Calculate (use a hand calculator) the appropriate powers of 3, sandwiching powers of two, relevant exponents, etc., to complete table 2.4.

Table 2.4

n	3^n	2^n	m	Limiting values of x
1	3	2	1	1/1 and 2/1
2	9	4	3	
3				
4				
5				
6				
7				
8				
9				
10				

Remember that if $2^m < 3^n < 2^{m+1}$ then

$$\frac{m}{n} < x < \frac{m+1}{n}.$$

38. Use the same procedure to estimate $L(5)$ and $L(7)$.
39. Suppose you change condition (2.9) to

$$L(10) = 1, \tag{2.9a}$$

Re-do exercises 37 and 38, on the basis of assumptions (2.7), (2.8), and (2.9a).
40. Let us compare our best approximations in the above exercises (use the decimal expressions of the approximations) by completing table 2.5.

Table 2.5

a	$L(a)$ using	
a	(2.9)	(2.9a)
2	1	
3		
5		
7		
10		1

How can you fill in the blank space in the last row with hardly any additional work? Does there seem to be any simple relation between the two columns? If you have a plausible conjecture, calculate a row corresponding to $a = 11$ and test your idea.

More efficient ways to compute $L(a)$
The method we sketched for estimating $L(3)$ works well theoretically. However, in order to calculate $L(3)$ to within an error of less than 0.01, we would need to compute 3^{100}, which is a number of 48 decimal digits. Use your hand calculator to compute the corresponding value of $L(3)$. Is the result you get really within an error of less than 0.01? Why?
There are better ways to compute $L(a)$, and we will now develop one of them. In chapter 3 we will suggest still more efficient procedures.
We shall outline the new method in a sequence of inequalities:

$2^1 < 3 < 2^2$, $1 < x < 2$;
$1 < 3/2 < 2$, $0 < x - 1 < 1$;
$(3/2)^1 < 2 < (3/2)^2$, $x - 1 < 1 < 2(x-1)$;
$1 < 4/3 < 3/2$, $0 < 2 - x < x - 1$;

$$4/3 < 3/2 < (4/3)^2, \qquad\qquad 2 - x < x - 1 < 2(2 - x);$$
$$1 < 9/8 < 4/3, \qquad\qquad 0 < 2x - 3 < 2 - x;$$
$$(9/8)^2 < 4/3 < (9/8)^3, \qquad\qquad 2(2x - 3) < 2 - x < 3(2x - 3);$$
$$1 < 256/243 < 9/8, \qquad\qquad 0 < 8 - 5x < 2x - 3;$$
etc.

On the left-hand side in each line, we have an inequality involving powers of 2 and 3. On the right, we have the inequality obtained from it by using (2.8), for example,

$$L\left[\left(\frac{9}{8}\right)^2\right] = 2L\left(\frac{9}{8}\right) = 2\left[L(9) - L(8)\right] = 2(2x - 3).$$

Exercises

41. Find the value of m in the last line of the above list of inequalities. You may find a hand calculator useful.
42. What estimate for x do you obtain from $0 < 8 - 5x$? What does the inequality $8 - 5x < 2x - 3$ tell you? How does the work in this method compare with the work in the previous method (p. 48)?
43. Do exercises 38, 39 by the present method. Compare your results also with exercise 40.

Suppose $L_{10}(x)$ is the function determined by conditions (2.7), (2.8), and (2.9a), and $L_2(x)$ is the one determined by (2.7)–(2.9). Let

$$F(x) = \frac{L_{10}(x)}{L_{10}(2)}.$$

We can easily check that $F(x)$ satisfies (2.7)–(2.9). For example,

$$F(ax) = \frac{L_{10}(ax)}{L_{10}(2)} = \frac{L_{10}(a) + L_{10}(x)}{L_{10}(2)}$$

$$= \frac{L_{10}(a)}{L_{10}(2)} + \frac{L_{10}(x)}{L_{10}(2)}$$

$$= F(a) + F(x).$$

Hence, by our uniqueness result, the function F coincides with L_2:

$$F(x) = L_2(x) \text{ for all } x \geqslant 1,$$

or

$$L_{10}(x) = L_{10}(2) L_2(x) \text{ for all } x > 1. \qquad (2.10)$$

Thus if we know the one number $L_{10}(2)$, then we can transform a table of each of these functions into a table of the other. If we apply (2.10) to $x = 10$ and use (2.9a), we obtain

$L_{10}(2) L_2(10) = 1.$

This shows us how to obtain either of the numbers $L_2(10)$ and $L_{10}(2)$ from the other.

Remarks on approximations and computations

We attach considerable importance to computations. In fact we constantly use approximations obtained by series of inequalities. Mathematically speaking, our main tools are Cauchy sequences. (We remind you that a sequence $\{x_n\}$ in a metric space with metric ρ is Cauchy if, given $\epsilon > 0$, there exists an integer N such that $\rho(x_p, x_q) \leqslant \epsilon$ whenever $p, q \geqslant N$.) In a complete metric space a Cauchy sequence is convergent, so if we work with real numbers Cauchy sequences will converge.

Theoretically, therefore, our way of working with series of inequalities provides the background for introducing the students to the use of ϵ in analysis, to notions of limit and of convergence, and to the properties of real numbers.

Our approach emphasizes the geometric view of nested intervals, which is intuitively understandable, and gives the teacher the possibility of suiting the degree of rigor of his way of handling the material to the level of the class.

It is also important that students be exposed early to the idea that most numbers, especially those coming from measurements, are not known exactly, and that we must ordinarily describe them by means of inequalities. This should be emphasized frequently at every level of school teaching. For example, one should teach that the statement that a length is 2.34 cm is an abbreviation for an inequality:

2.335 cm $<$ length $<$ 2.345 cm.

In computations with approximations one should be aware of the magnitude of the errors in the result due to the errors in the data. For example, the area of the rectangle in fig. 2.10 might by anything between 3.630 925 and 3.669 925, so that only the first two decimal places are meaningful at all.

Fig. 2.10

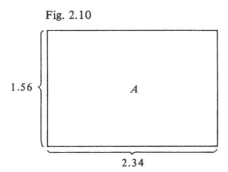

1.56

A

2.34

Log and log-log paper

Although the slide rule is now obsolete for practical calculations, the scale which we constructed in section 2.5 still has important uses. Suppose two variables y and x are related by the equation

$$y = 3 \times 5^x.$$

If x increases by an amount h, then the ratio of the new value of y to the old one is

$$\frac{3 \times 5^{x+h}}{3 \times 5^x} = 5^h,$$

which is independent of x. Thus equal differences in the values of x correspond to equal ratios in the values of y. If we graph the above relation, not on ordinary graph paper but on paper which uses the scale we constructed in section 2.5, on the y-axis and the usual scale on the x-axis, we get a straight line as in fig. 2.11.

Fig. 2.11

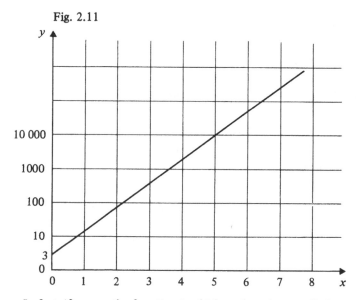

In fact, if we use the function L which we have just studied, we obtain

$$L_2(y) = L_2(3 \times 5^x) = L_2(3) + L_2(5^x) = L_2(3) + x L_2(5).$$

Thus we get a line which crosses the y-axis at the 3-point on the new scale. Its slope is the coefficient $L_2(5)$ of x. An increase of 1 in x corresponds to an increase of $L_2(5)$ in $L_2(y)$, which corresponds to an increase by ratio of 5 on the new scale.

Similarly, suppose that y and x are related by the equation

$$y = 3x^5.$$

If we double x, the ratio of the corresponding values of y is

$$\frac{3 \times (2x)^5}{3x^5} = 2^5,$$

which is independent of y. In general, equal ratios in the values of x correspond to equal ratios in the values of y. Thus if we graph this relation on paper which uses the new scale of section 2.5 on both axes, then we get a straight line as in fig. 2.12.

Fig. 2.12

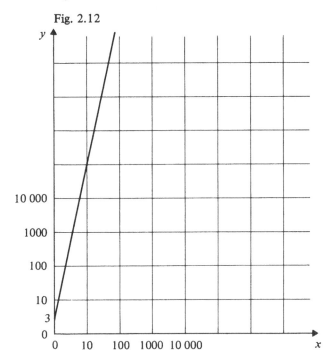

In fact, if we use the function L, we obtain

$$L_2(y) = L_2(3x^5) = L_2(3) + 5L_2(x).$$

This is the equation of the straight line which crosses the y-axis at the 3-point on the new scale, and which has the slope 5, as measured on the old scale.

Graph paper drawn to the scales used in fig. 2.11 is called log paper, and graph paper drawn to the scales used in fig. 2.12 is called log–log paper.

Another, maybe simpler, case where it is worth while to use log–log paper is when x and y are related by the equation

$$xy = c,$$

where c is a constant. Indeed, using the function L we get

$$L_2(x) + L_2(y) = L_2(c),$$

and drawn with log–log scales the corresponding graph is a straight line (see fig. 2.13).

Fig. 2.13

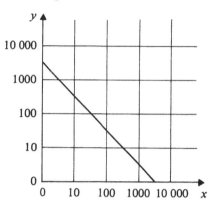

3

Exponentials

This chapter centers on the notion of 'rate of change', going from the average rate to the instantaneous one. This is done by examining different problems of growth. Our mathematical models lead us to geometric progressions, exponential functions, difference and differential equations.

We start with two different real-life situations: the struggle for life, and radioactive decay, both of which lead to the same mathematical model. Both units are written as texts for students.

In section 3.3 we show how to present geometric and arithmetic progressions in a meaningful way. Arithmetic progressions of a higher order are treated at the junior high school level.

Difference equations versus differential equations are considered in section 3.4. We consider the study of this dual approach to mathematical models a must for college-level mathematics. In our text we examine both points of view, the relationship between them, and their relationship with exponential, or logarithmic, functions. We also explain, at some length, how to approach exact computations, a topic which is hard to find treated in textbooks. Pointers are given on how to study a differential equation directly. The chapter ends with some more general remarks on mathematical models.

3.1   The mathematical theory of the struggle for life

Historical note
During World War 1, fishing in the Adriatic Sea (locate this on a map) was interrupted. After the war, when the Italians resumed fishing they were surprised to find fewer fish of the kind they had been catching than there had been before. They thought, of course, that since they had not been catching these fish for four years there would be many more of them.

One of the executives of the Italian fishing industry was the son-in-law of

Professor Vito Volterra, one of the greatest of Italian mathematicians. The executive asked Volterra whether he could find an explanation. He worked out a mathematical theory of the struggle for life, which you will be able to read as soon as you learn French and calculus (V. Volterra, *Leçons sur la théorie mathématique de la lutte pour la vie*, Gauthier-Villars, Paris, 1931). We shall try in this chapter and chapter 6 to explain some of his main ideas in terms of high school algebra.

If you wish to pursue these problems further you may read the book by Professor A. J. Lotka of Massachusetts Institute of Technology, *Elements of mathematical biology* (Dover Publications, New York, 1956). You may also wish to explore the work of Professor Sewall Wright, of Northwestern University, on the quantitative theory of evolution (S. Wright, 'Statistical genetics and evolution', *Bull. Am. Math. Soc.*, 48 (1942), 223–46; S. Wright, 'The genetical structure of populations', *Ann. Eugenics*, 15 (1951), 323-54).

Growth of a single population: simplest version

Suppose we consider a population of bacteria in a culture dish or in your blood stream, and count the population every day. Let $x(t)$ be the population t days after we begin the experiment, so that $x(0)$ is the initial population, $x(1)$ is the population after one day, and so on.

Suppose we obtain the results shown in table 3.1. The change in the population during the first day is the population at the end of one day minus the population at the start, or $x(1) - x(0)$. Thus we find that the change during the first day is

$$x(1) - x(0) = 1\,010\,000 - 1\,000\,000 = 10\,000.$$

Similarly, the change during the second day is

$$x(2) - x(1) = 1\,020\,100 - 1\,010\,000 = 10\,100.$$

Fill out the rest of table 3.1.

During each day any particular bacterium has a certain chance of reproducing and a certain chance of dying. Say that the chance, or probability, of

Table 3.1

t	$x(t)$	Change in population
0	1 000 000	10 000
1	1 010 000	10 100
2	1 020 100	
3	1 030 301	
4	1 040 604	
5	1 050 010	

reproducing during the period of one day is 0.03, that is, on average three bacteria out of 100 reproduce during one day. Suppose also that the chance of any particular bacterium dying during this period is 0.02. Then the excess of births over deaths is $3 - 2 = 1$ out of 100 per day. In other words, the relative rate of growth is 1 per cent, or 0.01 per day.

We could state this result in the following way. The relative rate of growth during the first day is the proportion of the change in the population to the whole population:

$$\frac{x(1) - x(0)}{x(0)} = \frac{1\,010\,000 - 1\,000\,000}{1\,000\,000}$$

Work out this ratio. What is the relative rate of growth of the population during the first day?

Exercises

1. Compute the relative rate of growth of the population in table 3.1 for each day.

    Day   1   2   3   4   5
    Relative rate
    of growth

    Carry out your computations to two decimal places. What do you notice?
2. Assuming that the relative rate of growth remains constant, predict the value of $x(6)$.
3. Solve the equation
    $$\frac{x(7) - x(6)}{x(6)} = 0.01.$$
4. Solve the equation
    $$\frac{x(t + 1) - x(t)}{x(t)} = r$$
    for $x(t + 1)$ as the unknown.

The difference equation

As a first approach, on the basis of a great deal of experimentation, we shall assume that the relative rate of growth of the population is a constant. If this constant is 0.01, then we obtain the following series of equations:

$$\frac{x(1) - x(0)}{x(0)} = 0.01,$$

$$\frac{x(2) - x(1)}{x(1)} = 0.01,$$

$$\frac{x(3) - x(2)}{x(2)} = 0.01,$$

and, in general,

$$\frac{x(t + 1) - x(t)}{x(t)} = 0.01.$$

Solve these equations for $x(1), x(2), x(3), \ldots$, and in general for $x(t + 1)$ in terms of $x(t)$:

$$x(t + 1) = (\quad) x(t).$$

Fill in the missing values in this and all subsequent such expressions. Now you have an expression for $x(2)$ in terms of $x(1)$, and an expression for $x(1)$ in terms of $x(0)$. Obtain a formula for $x(2)$ in terms of $x(0)$:

$$x(2) = (\quad) x(0).$$

In the same way, find expressions for $x(3), x(4)$, and $x(5)$ in terms of $x(0)$. Try to guess a formula for $x(t)$ in terms of $x(0)$:

$$x(t) = (\quad) x(0).$$

Our next step is to solve the problem for any relative rate of growth. Let us assume that this rate of growth is a constant, r. Then we have the equations

$$\frac{x(1) - x(0)}{x(0)} = r,$$

or

$$x(1) - x(0) = rx(0),$$
$$x(2) - x(1) = rx(1), \ldots,$$
$$x(t + 1) - x(t) = rx(t).$$

Solve these equations for $x(1), x(2), \ldots, x(t + 1)$, respectively, and obtain an expression for each day's population in terms of the previous day's population:

$$x(1) = (\quad) x(0), \tag{3.1}$$
$$x(2) = (\quad) x(1), \ldots, \tag{3.2}$$
$$x(t + 1) = (\quad) x(t). \tag{3.3}$$

As a check, compare with table 3.1, where $x(0) = 1\,000\,000$ and $r = 0.01$, and also with the previous results of this section.

Now express $x(2), x(3), \ldots, x(t)$ in terms of $x(0)$. Substitute in equation (3.2) the value of $x(1)$ from equation (3.1). Write down the expression for $x(3)$ in terms of $x(2)$. Substitute in this equation for $x(2)$ its expression in terms of $x(0)$. Guess at a formula for $x(t)$:

$$x(t) = (\quad)^? x(0),$$

and check by substituting $t = 0$, $t = 1$, $t = 2$, and $t = 3$.

So far we have imagined that the population is counted every day. We can also consider what happens if we use a different time interval. Suppose we count (or estimate) the population every h days. The number h might be 7 (weekly observations), 1/24 (hourly observations), or any other number we choose. Let us assume that the relative change in population *per unit time* is a fixed number r.

Then our observations are made at the times

$$t = 0, h, 2h, 3h, \ldots, nh, \ldots,$$

and the observed populations are

$$x = x(0), x(h), x(2h), x(3h), \ldots, x(nh), \ldots$$

The changes in population are

$$x(h) - x(0), x(2h) - x(h), \ldots, x[(n + 1)h] - x(nh),$$

and the relative changes are

$$\frac{x(h) - x(0)}{x(0)}, \quad \frac{x(2h) - x(h)}{x(h)}, \text{ etc.}$$

The *relative rate of change* during the first time interval is

$$\frac{\text{relative change}}{\text{length of time interval}} = \frac{x(h) - x(0)}{x(0)} \cdot \frac{1}{h}$$

$$= \frac{x(h) - x(0)}{hx(0)}.$$

By our assumption, this must be equal to the given constant r:

$$\frac{x(h) - x(0)}{hx(0)} = r,$$

and we obtain

$$x(h) - x(0) = rhx(0).$$

Write down the equation for each of the other time intervals. In particular, the equation for the $(n + 1)$th time interval is

$$x[(n + 1)h] - x(nh) = rhx(nh).$$

Solve these equations to express $x(h)$ in terms of $x(0)$, $x(2h)$ in terms of $x(h), \ldots,$ and $x[(n + 1)h]$ in terms of $x(nh)$:

$$x[(n + 1)h] = (\quad) x(nh).$$

Now solve these equations to obtain expressions of $x(h), x(2h), \ldots, x(nh)$, in terms of $x(0)$:

$$x(nh) = (\quad)^? x(0).$$

Remember now that $t = nh$. Solve this equation for n in terms of t, and obtain the following formula for $x(t)$ in terms of t and $x(0)$:

$$x(t) = C^t\, x(0).$$

Give a formula for the constant C in terms of r and h:

$$C = (\quad).$$

Exercises

5. Check your result by computing C when $r = 0.01$ and $h = 1$.
6. Complete table 3.2 of the values of C for various values of r and h.

Table 3.2

r	h	C	r	h	C
1	1		0.5	0.005	
1	0.5		2	1	
1	0.1		2	0.5	
1	0.01		2	0.1	
1	0.001		2	0.01	
0.5	1		2	0.001	
0.5	0.5				
0.5	0.1				
0.5	0.01				

7. Let $C(r,h)$ be the value of C for given values of r and h. Compute $|C(1, 0.001)|^2$, $|C(0.5, 0.001)|^2$ and compare with $C(2, 0.001)$, $C(1, 0.001)$ respectively.

8. (a) Prove the inequality

$$(1 + r)^2 > 1 + 2r \text{ if } r > 0.$$

(b) Prove that if $r > 0$ and

$$(1 + r)^n > 1 + nr.$$

then

$$(1 + r)^{n + 1} > 1 + (n + 1)r.$$

(c) Prove that

$$(1.000\,001)^{1\,000\,000\,000} > 1001.$$

(d) Find a number n such that

$$(1.000\,001)^n > 1\,000\,000.$$

Exercise 8 shows that if a population grows according to the law discussed above then it ultimately becomes larger than any number you may choose. This, of course, does not in fact happen. Any bacterial population will be limited in size, either by lack of room or lack of food. This means that our mathematical model of the growth of a population is not good enough. It might be appropriate for a while, but not in the long run. In chapter 6 we will reconsider the problem and try to find a better model.

3.2 Radioactive decay

The law of decay

Suppose you have a rock containing uranium. If each day you were to measure the amount of uranium in your sample, you would find the amount decreasing daily as the uranium changed into lead. Since the decay of uranium is a slow process, we shall consider instead an artificial element produced in the laboratory, which obeys the same law of decay as uranium, but which decays much faster.

A physicist produced a small amount of ^{52}Fe, a radioactive isotope of iron (Fe), in a one-gram sample of iron, and measured the amount of ^{52}Fe in the sample every hour. Table 3.3 shows the amount of ^{52}Fe he found at various times after its manufacture. In the third column we have left space for you to fill

Table 3.3

t (hours)	x (mass in units of 10^{-20} grams)	Δx	$\Delta x/x$
0	2.50		
1	2.29		
2	2.09		
3	1.93		
4	1.77		
5	1.63		
6	1.48		

in the change in the amount of ^{52}Fe from one hour to the next. For example, during the first hour the change is

new value — old value = 2.29 – 2.50 = – 0.21 ($\times$ 10^{-20} grams).

We write this number in the first row of the third column. Similarly, the next entry in the third column is 2.09 – 2.29 = –0.20. Fill in the rest yourself. The changes are all negative since the amount of ^{52}Fe is decreasing. It is convenient to introduce a symbol for the successive changes in x:

Δx = change in x.

Δ is the Greek letter delta, so Δx may be read as 'delta x'. The head of the third column of table 3.3 thus means 'change in x'.

It is interesting to know what proportion of the ^{52}Fe disintegrates each hour, so we have made a fourth column for $\Delta x/x$. Compute these ratios, and write them in the fourth column. The first entry is

$$\frac{-0.21}{2.50} = -0.084.$$

Fill in the rest yourself.

■   Definition The value of $\Delta x/x$ each hour is the *relative change of x* during that hour.

What do you notice about the values of $\Delta x/x$? Can you predict the next few numbers in that column? Can you tell from this what the next number in the Δx-column will be? How can you use this to predict the next number in the x-column? Repeat this process and predict the next few lines in the table. Compare your answers with the experimental results shown in table 3.4. How well do your predictions agree with the experimental results? Can

Table 3.4

t	x	Δx	$\Delta x/x$
7	1.37		
8	1.25		
9	1.14		
10	1.05		
11	0.97		
12	0.89		

you predict when there will be less than 0.01×10^{-20} grams of ^{52}Fe in the sample? Why do we not find ^{52}Fe in nature?

Table 3.4 agrees, to good approximation, with the following law:

The relative change of x per hour is *constant*.

This is another way of saying that the numbers in the last column are constant. We can also state this law in the form of an equation:

$$\frac{\Delta x}{x} = c,$$

where c is a certain fixed number. We cannot expect such a law to fit the observed data exactly. Which value of c do you think fits the last column best?

^{52}Fe illustrates the general law of radioactive decay, which applies to all radioactive substances:

The relative change in the amount of a substance during time intervals of a given length is a constant.

A theory which gives a good explanation of this law is that any particular atom has a certain definite *probability* of disintegrating within a given interval of time. For example, suppose that the probability of any ^{52}Fe atom disintegrating during an hour is 0.083. Then in our sample, which contains a very large number of ^{52}Fe atoms (of the order of 10^{12}), it is almost certain that about 0.083 of the sample will disintegrate during any one hour. Thus the decrease of ^{52}Fe during one hour would be about 0.083 of the amount present at the start of the hour. This would agree well with the data in our table. We can express the law more concisely in terms of the concept of relative rate of change.

■   Definition The *relative rate of change* of x is the relative change per unit time, or relative rate of change of x

$$= \frac{\text{relative change of } x}{\text{length of time interval}} .$$

In our example, the time intervals all have the length of one hour, so that the numbers in our calculations are left unchanged.

In exercises 9–13 we shall use $x(t)$ to stand for the value of x at the time t, that is, t hours after the start of the experiment. We read x_3 as 'x sub 3'. Be careful not to confuse x_3 with x^3, which means x to the third power.

Exercises

9. Take for c the average of the numbers in the fourth column of table 3.4. Predict the values of $x(13)$, $x(14)$, and $x(15)$. Round off your results to two decimal places.
10. Make a graph showing the relation between t and x in the table 3.4.
11. Suppose that the measurements were made every half day, but that the same relative rate of change was observed. Make a table showing the values of x for $t = 0, 0.5, 1, 1.5, 2, \ldots, 5$.
12. Make a graph showing the relation in exercise 11.
13. The following table shows the amounts of ^{103}Pd (a radioactive isotope of palladium) on successive days.

t (days)	0	1	2	3	4	5
x (mass in units of 10^{-10} grams)	2.84	2.79	2.64	2.52	2.41	2.36

     Does the law of radioactivity fit this table approximately? What value of c fits these data best? Predict the values of $x(6)$ and $x(7)$; the experimental values were 2.26 and 2.18 respectively. How good were your predictions?

Geometric progressions

We now wish to analyze the mathematical model of the law of radioactive decay. We wish to derive a few consequences; some of these may lead to the prediction of new experimental results.

It will be convenient in this section to write $x_0, x_1, \ldots, x_n$ for the successive values of x, so that $x(0) = x_0, x(1) = x_1, \ldots$ Thus the change during the first time interval is $x_1 - x_0$, and the relative change during this time interval is $(x_1 - x_0)/x_0$.

Suppose, for the sake of simplicity, that we measure x every hour, and that the relative rate of change is -0.1 per hour. We can express this as

$$\frac{x_1 - x_0}{x_0} = \frac{x_2 - x_1}{x_1} = \ldots = -0.1 .$$

We can solve the equation

$$\frac{x_1 - x_0}{x_0} = -0.1$$

for x_1 in terms of x_0:

$$x_1 = (\quad)x_0. \tag{3.4}$$

Fill in the missing terms in this and any subsequent such expressions. Similarly, you can solve the second equation

$$\frac{x_2 - x_1}{x_1} = -0.1$$

for x_2 in terms of x_1:

$$x_2 = (\quad)x_1, \tag{3.5}$$

and so on. Continuing in this way, what do you notice about the ratios

$$\frac{x_1}{x_0}, \frac{x_2}{x_1}, \frac{x_3}{x_2} ?$$

■ Definition A sequence of numbers such that the ratio of any two consecutive numbers in the sequence is a constant is called a *geometric progression*. The constant is called the *common ratio* of the progression.

Do the numbers $x_0, x_1, \ldots$, form a geometric progression? If so, what is the common ratio? Equations (3.4) and (3.5) tell us how to predict one hour ahead. How can we predict n hours ahead, that is, x_n from x_0?

In equation (3.5) you can substitute the value of x_1 from equation (3.4). You obtain

$$x_2 = (\quad)x_0.$$

Express x_3 in terms of x_0, x_4 in terms of x_0. Give a formula expressing x_n in terms of x_0:

$$x_n = (\quad) x_0.$$

Exercises

14. Suppose that the relative change per day is -0.96, and let $x_0 = 2.84$.
    (a) Find the values of $x_1/x_0, x_2/x_1, x_3/x_2, x_4/x_3$.
    (b) Find a formula for x_n in terms of x_0.
    (c) How well does your formula in (b) fit the data on ^{103}Pd given in exercise 13?

15. Suppose that the relative rate of change of x is -0.1 per hour, but that the observations are made every half hour. Let $x_0, x_1, x_2, \ldots,$ be the successive measurements of x. Thus we have

$$\frac{x_1 - x_0}{x_0} \Big/ \frac{1}{2} = -0.1,$$

and so on.
    (a) Find a formula expressing x_n in terms of x_0.
    (b) Assume $x_0 = 100$ and calculate $x_1, x_2, \ldots, x_8$ using the formula found in (a).
    (c) Suppose now that the observations are only made every hour. Let $x_0^*, x_1^*, x_2^*, \ldots,$ be the successive measurements. Obviously $x_0^* = x_0$. Compute $x_1^*, x_2^*, x_3^*,$ and x_4^*. Compare x_n^* with the value of x_{2n} from (b).

Solution for any radioactive material
    Suppose that we have a sample containing some radioactive material and measure the amount of the radioactive material every h days, that is, we measure the amount at the times

$$t = 0, h, 2h, 3h, \ldots, nh, \ldots$$

We find the amounts

$$x = x_0, x_1, x_2, x_3, \ldots, x_n, \ldots$$

in the sample. What is the relative change in x during the first h days? What is the relative rate of change in x per day?
    By the law of radioactive decay, this relative rate of change in x per day must be a constant. In the case of ^{52}Fe this change is about -0.083 per hour, or $-0.083/(1/24) = -1.99$ per day, while in the case of ^{103}Pd it is about -0.04 per day. The constant is negative since the amount is decreasing.
    In the general case, the constant can be set equal to $-k$, where k is a positive number. We obtain the equation

$$\frac{x_1 - x_0}{x_0} / h = -k, \tag{3.6}$$

or

$$\frac{x_1 - x_0}{hx_0} = -k.$$

If we multiply both sides of this equation by hx_0, we obtain

$$x_1 - x_0 = -khx_0.$$

We can now add x_0 to both sides, and factorize the right-hand side. We have now solved equation (3.6) for x_1 and have obtained

$$x_1 = (1 - kh)x_0. \tag{3.7}$$

Set up the equation describing what happens during the next h days, and solve this equation for x_2 as the unknown. In this equation, substitute the value of x_1 in terms of x_0, and thus express x_2 in terms of x_0. In general, the equation

$$\frac{x_{n+1} - x_n}{hx_n} = -k$$

describes what happens during the $(n + 1)$th time interval of h days. Solve this equation for x_{n+1} in terms of x_n:

$$x_{n+1} = (\quad)x_n.$$

If you know x_n, how can you calculate x_{n+1}?

Exercises

16. Express x_3 in terms of x_2.
17. Express x_3 in terms of x_0.
18. Express x_4 in terms of x_0.
19. Express x_n in terms of x_0.
20. Let $x(t)$ denote the amount at the time t. Give a formula of the form

    $$x(t) = C^t x_0,$$

    where C is a certain constant expressing $x(t)$ in terms of t. (Hint: we already have such an expression for $x(t)$ when $t = nh$ in section 3.1. Can n be eliminated from that expression?)
21. Take $k = 0.1$. Compute C in exercises 11–13 for $h = 1, 0.5, 0.1, 0.01, 0.001$. What do you notice about the values of C?
22. The time T at which half of the radioactive substance has disintegrated is called the half-life. In other words, the *half-life* is the solution T of the equation

$$C^T x_0 = \frac{x_0}{2}.$$

Obtain a formula for T in terms of k and h. (Hint: take logs.)
23. It is sometimes easier to measure T than k. Find a formula expressing k in terms of T and h. For $T = 10$, compute k for $h = 1, 0.5, 0.1, 0.01, 0.001$. What do you notice about the value of k?

Exercises 21-23 are for students who have studied logarithms.

The exponential function
The function has so far appeared in section 3.1 and this section only in the exercises; in section 3.4 we will look at the function in its theoretical aspects. The easiest way to start studying the function at high school level is by looking at a specific example, for instance

$$y = 2^x,$$

and asking the students to draw the corresponding graph, taking integer values of x first and then successively finer divisions of the resulting intervals between integer values of x.

Tables similar to those we made before could successfully be used. Tables 3.5 and 3.6 show how these could be started. The tables can be computed

Table 3.5

x	y	Δy	$\Delta y/y$
0	1	0.4142	0.4142
0.5	1.4142	0.5858	0.4142
1	2		

Table 3.6

x	y	Δy	$\Delta y/y$
0	1		
0.25			
0.5	1.4142		
0.75			
1	2		

using the square root operation, especially if hand or desk calculators are available. Questions that should be investigated are:
(*a*) What is the new constant value of $\Delta y/y$ and why?

(b) From a certain row onwards, the numbers in the columns y and
Δy are just twice numbers in those columns in a preceding row. Why?

(c) Can you go on with similar procedures and come as close as you
wish to any real positive value of x?

3.3 Malthus: an elementary view

The birth rate

We can treat some aspects of the exponential function at a more
basic level in connection with a discussion of the famous theory of Malthus
regarding population growth and food supply. Since these are now matters of
considerable public interest, one might encourage the students, for motivation purposes, to bring into class newspaper items concerning the 'population explosion' problem of supply of food and other natural resources, and
its effects on the policies of various countries, such as campaigns for family
planning.

Suppose the birth rate of a country is 20 per thousand per year. This
means that for every 1000 people an average of 20 babies will be born during
a year. In other words, the *increase* of the population will be

$$\frac{20}{1000} = 0.02 = 2\%$$

each year, ignoring deaths.

If there are 1 000 000 people in the country at the start of this year, then
the increase during the year will be

$$0.02 \times 1\,000\,000 = 20\,000,$$

so that the population at the beginning of the next year will be

$$1\,000\,000 + 20\,000 = 1\,020\,000.$$

We can continue this process and make a table of the population P versus
the time t, measured in years from now. Table 3.7 shows that in ten years
the population increases by 218 997, or about 22% of the original population.
We have made a crude extrapolation in table 3.8, which shows that the population will be more than double in forty years.

Exercises

24. Divide the class into groups of about five students, and assign to
each a birth rate to investigate. For various initial populations they
should make tables like tables 3.7 and 3.8. For each birth rate,
estimate the time of doubling.

Table 3.7

Time t	Population P	Annual increase in P
0	1 000 000	20 000
1	1 020 000	20 400
2	1 040 400	20 808
3	1 061 208	21 224
4	1 082 432	21 649
5	1 104 081	22 082
6	1 126 163	22 524
7	1 148 687	22 974
8	1 171 661	23 434
9	1 195 095	23 902
10	1 218 997	

Table 3.8

t	P	Approximate increase in P per decade
0	1 000 000	220 000
10	1 220 000	244 000
20	1 464 000	292 800
30	1 756 800	

25. What are some of the ways in which the above model is a simplification of reality? For example, what tacit assumption did we make
in calculating the increase during the fourth year? How would the
model be modified if we assumed a birth rate of 50 per thousand
per year and a death rate of 30 per thousand per year? Who has
babies?

The food supply

Malthus also assumed that the food supply increases by a fixed
amount each year. If we assume a supply of 2 000 000 metric tons at the
start and an increase of 30 000 metric tons per year, we can calculate the
increase in food supply, as in table 3.9.

We have combined the population and food supply tables in table 3.10. It
is interesting to add a column for F/P, the food supply per person. We see
that in ten years the food supply per person sinks from 2.0 to less than 1.74
metric tons per person. Estimate F/P at the end of thirty and sixty years.
Suppose that the subsistence level is 0.1 metric tons of food per person per
year. How long will it be before there is a famine?

Table 3.9

Time t	Food supply F
0	2 000 000
1	2 030 000
2	2 060 000
3	2 090 000
4	2 120 000
5	2 150 000
6	2 180 000
7	2 210 000
8	2 240 000
9	2 270 000
10	2 300 000

Table 3.10

t	P	F	F/P
0	1 000 000	2 000 000	2.0
1			
2			
3			
4			
5			
6			
7			
8			
9			
10	1 321 215	2 300 000	1.74

Exercises

26. Have each group try various initial food supplies and rates of increase. Let them make comparison tables in each case. Does F/P ever begin to decrease? When? Does it always become small ultimately? How does the time it takes for F/P to become small, say less than 0.1, depend on the initial values and the rates of increase?

27. If F/P becomes lower than the subsistence level, what will happen to the death rate? Will P continue to increase at a constant ratio? Can the above model be completely realistic?

The successive values of P in our table have a constant ratio of 1.02:

$$\frac{1\,020\,000}{1\,000\,000} = \frac{1\,040\,040}{1\,020\,000} = 1.02.$$

Such a sequence is called a *geometric* progression. The number 1.02 is called the *common ratio* in this progression. The successive values of F have a constant difference of 30 000:

$$2\,030\,000 - 2\,000\,000 = 2\,060\,000 - 2\,030\,000$$
$$= 30\,000\,.$$

Such a progression is called an *arithmetic* progression, and this number 30 000 is called its *common difference*. It seems from the above calculations that a geometric progression with ratio greater than 1.0 grows faster than any arithmetic progression.

Exercises

28. Let $P(t)$ be the population at time t in the above example, and $F(t)$ be the food supply at time t. Find formulas for $P(t)$ and $F(t)$ in terms of t. Generalize to an arbitrary initial population $P(0)$, common ratio r, initial food supply $F(0)$, and common difference d.

29. Let $R(t) = F(t)/P(t)$ be the food supply per person. In the above example, find a good formula for $R(t + 1)/R(t)$. When is this ratio less than 1.0? When is it less than 0.99? Is it ever less than 0.98?

30. Discuss the questions of exercise 28 in the general case.

31. It may be more realistic to assume that the amount of food consumed each year is proportional to the population, that is,

    amount of food consumed $= aP$,

    where a is a certain constant. With a fixed agricultural technology and amount of farm land, the formula

    amount of food produced $= bP^k$

    works fairly well where b and k are constants and k is less than 1.0. This gives us the relation

    $F(t + 1) - F(t) =$ change in food supply
    $\qquad\qquad\qquad = $ amount produced $-$ amount consumed
    $\qquad\qquad\qquad = b[P(t)]^k - aP(t).$

    Assume $a = 5, k = 0.7, b = 530$. Calculate a table of $F(t)$ versus t using the population figure from our example and $F(0) = 2\,000\,000$ as before. Make a table of $t, P, F,$ and F/P. How does this table compare with the previous table? Now when does F/P begin to decrease? When does it fall below 0.1? For which populations, if any, does the food supply decrease?

32. Which is more realistic, the model in exercise 31 or Malthus's model? Which is more pessimistic? In what ways does this model fail to fit reality?

33. Table 3.11 gives the census figures for the United States at ten-year intervals.

Table 3.11

Year	Population
1800	5 308 483
1810	7 239 881
1820	9 638 453
1830	12 866 020
1840	17 069 453
1850	23 191 876
1860	31 441 321
1870	38 558 371
1880	50 155 783
1890	62 622 250
1900	75 994 575
1910	91 976 266
1920	105 710 620
1930	123 203 000

Analyze the data. Does the model we used in exercise 31 give a good approximation? For the table as a whole? For parts of the table? Why?

Other progressions

We have studied two types of progression, the arithmetic and the geometric, which can be approached by arithmetic and by elementary algebra. There is another type of progression which occurs often and which also lends itself to simple treatment. It is sometimes called an arithmetic progression of a higher order.

Galileo found an example of this type of progression when he studied a ball rolling down an inclined plane under the influence of gravity (fig. 3.1).

Fig. 3.1

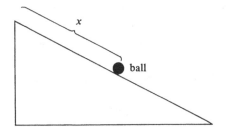

He obtained data similar to those in table 3.12 (we are not giving his actual numbers). During the first second the ball rolled $10 - 0 = 10$ cm. How far does it travel during the second, third, and fourth seconds? Compute the third column in table 3.12, Δx, the change in x. What do you notice about the sequence in the third column? Is it of a type you have met before? Can you predict the next number in the third column? Can you predict the next number in the second column?

Table 3.12

t (seconds)	x (centimeters)	Δx	$\Delta(\Delta x)$
0	0		
1	10		
2	22		
3	36		
4	52		
5			

The changes in the successive changes form one of the simplest types of sequence. We have made a fourth column in table 3.12 for the *changes* in Δx, labeled '$\Delta(\Delta x)$'. Calculate the sequence in the fourth column. If all the terms in the Δx column (Δ^1) were the same, x would be an arithmetic progression. Here all the terms in the $\Delta(\Delta x)$ column (Δ^2) are the same, so we call this an arithmetic progression of the *second order*.

Exercises

34. In table 3.13, which shows the area A of a square of side s, calculate the columns for ΔA and $\Delta(\Delta A)$. How is your result similar to table 3.12 for the rolling ball?

Table 3.13

s	A	ΔA	$\Delta(\Delta A)$
0	0		
1	1		
2	4		
3	9		
4			
5			
6			

35. The previous exercise suggests that if you can find a certain number k such that $y = x - kt^2$ then $\Delta(\Delta y) = 0$. Find k. What kind of progression will y be? Find a formula for y in terms of t. Find a formula for x in terms of t.

The method of these exercises can often be applied to find a polynomial which fits a given table of data at least approximately. For this purpose it is useful to make tables of the various powers of t and the differences. We can then use these in trying to find a polynomial which fits a given table.

Exercises

36. Calculate a table of t^3, where $t = 0, 1, 2, \ldots$, and calculate differences as above until you find a column of constant differences. Can you predict what will happen with a table of t^4? Check your prediction.

37. Find a polynomial $P(t) = at^3 + bt^2 + ct + d$ which fits this table:

t	0	1	2	3	4	5	6
x	0	1	5	14	30	55	91

38. Find a polynomial which approximately fits this table;

t	0	1	2	3	4	5	6
x	0	0.97	5.01	13.99	30.00	55.02	91.00

How good a fit can you get?

39. The boiling points of the straight chain hydrocarbons of the form $C_n H_{2n+2}$ are shown in table 3.14. What kind of progression do we get?

Table 3.14

n	BP (boiling point, °C)	ΔBP (Δ^1)	$\Delta(\Delta$BP) (Δ^2)
1	−161.7		
2	−88.6	73.1	−26.7
3	−42.2	46.4	−4.7
4	−0.5	41.7	−5.1
5	36.1	36.6	−4.0
6	68.7	32.6	−2.9
7	98.4	29.7	−2.5
8	125.6	27.2	−2.1
9	150.7	25.1	−1.8
10	174.0	23.3	

3.4   Differential and difference equations

The two viewpoints

In the discussions of radioactive decay and of the growth of a population, we were dealing with a function $x(t)$ which satisfies the equation

$$\frac{x(t+h) - x(t)}{h} = rx(t), \tag{3.8}$$

where r is a certain constant. This abstract mathematical equation is a common
model of these two apparently unrelated phenomena. We arrived at equation
(3.8) because we interpreted the laws of the phenomena in terms of average
rates of change. The *difference quotient* on the left is the average rate of
change of x during the time interval from t to $t + h$.

Suppose that the changes are occurring *continuously*. Then the time t
must be considered as a continuous variable, even though we only observed
the quantity x at the discrete times $0, h, 2h, 3h$, etc. As such, equation (3.8)
must be regarded as an approximation. We should really work with the
instantaneous rates of change, which are represented by the limit

$$\lim_{h \to 0} \frac{x(t+h) - x(t)}{h}, \tag{3.9}$$

obtained by averaging over ever smaller intervals. We recognize the limit in
(3.9) as the derivative $x'(t)$.

Hence we could formulate the mathematical model as

$$x'(t) = rx(t). \tag{3.10}$$

For small values of h the difference quotient in (3.8) should be close to the
derivative in (3.10), so we would expect the solutions to be close together
for such small values of h.

Equation (3.8) involves the differences $x(t+h) - x(t)$ of the values of the
unknown x at various times, so that equation is called a *difference equation*.
Equation (3.10) involves the derivative $x'(t)$ of the unknown function, so it
is called a *differential equation*.

The situation can also be exactly the opposite – the true model being the
difference equation and the approximation being given by a differential
equation. If, for example, $x(t)$ is the amount of money in your bank account
at time t, and you deposit the amount $x(0)$ at the time $t = 0$, then the changes
only occur at the ends of interest periods. If r is the annual rate of interest
and the bank compounds quarterly, then the model is equation (3.8) with
$h = 1/4$. Similarly, if the money is compounded daily, then $h = 1/365$ in an
ordinary year. (What is h for a leap year?) In this case, equation (3.8) is the

exact model and, if h is small, the solution of equation (3.10) may be expected to be a good approximation to the true solution.

Thus we see that sometimes the first of the above equations represents the true model, and the other an approximation, and sometimes the roles are reversed. In any particular case, we must be careful to know which situation occurs.

As we saw in section 3.1, equation (3.8) can be treated by means of elementary algebra. We found that the solution is

$$x(t) = x(0) \, C(h)^t \tag{3.11}$$

where

$$C(h) = (1 + rh)^{1/h} \tag{3.12}$$

On the other hand, equation (3.10) *essentially* involves calculus, and is therefore less elementary. But we expect that for small values of h the function in (3.11) should be close to the exact solution (3.10). This suggests that we begin by investigating how it behaves as h gets smaller.

Formula (3.11) shows that the behavior of the solution with varying h is completely determined by the behavior of $C(h)$. For given r this depends on the variable h, so it is simpler to study than $x(t)$, which also depends on t.

We may let $h = 1/2^n$ or $1/10^n$, with n increasing. Even though in the original biological or physical problem h is positive, it turns out to be useful to see also how $C(h)$ behaves for small negative values of h, such as $-1/2^n$ or $-1/10^n$ with large n.

Before doing some calculations, we must note that when a computer program 'solves' a differential equation, it in fact *replaces* the equation by a suitable difference equation, and works out calculations similar to the ones in the exercises.

So in practice it might sometimes be pointless to replace an easily understood difference equation by a more elegant, continuous, differential equation, only then to go back to the difference equation we had before when looking for numerical results.

Exercises

40. Choose a value for r and calculate a table for h and $C(h)$. Try $h = 1/2^n$ with $n = 0, 1, 2, 3, \ldots, 10$. Repeat with $h = -1/2^n$. You might also try $h = \pm 2/10^n$. Do you notice any trend? Compare your results with those of your classmates who chose other values of r. Did they find similar trends? Use your calculator.

41. Compare the formula for $C(2h)$ with the formula for $C(h)$. You will see that $C(2h) = A^{1/h}$ with $A = (\quad)$.

(a) Is $(1 + rh)^2 - A^2$ positive or negative? If $1 + 2rh > 0$, which is larger, $1 + rh$ or A?

(b) If $h > 0$ and $(1 + 2rh)$ is positive, which is larger, $C(h)$ or $C(2h)$?

(c) What about if h is negative?

(d) Does this explain what you found in exercise 40?

42. Compare $C(10h)$ with $C(h)$ in the same way. How does this agree with your calculations? It might be simpler to consider only the case where rh is positive.

43. Let $a_n = (1 + k)^n - 1 - kn$.

(a) For $n = 1$ or 2, is a_n positive, negative, or zero?

(b) Find a simple formula for $a_{n+1} - (1 + k)a_n$. Is this difference positive, negative, or zero?

(c) If $1 + k > 0$, can a_2 be negative? What about a_3? Could a_n be positive and a_{n+1} negative? Can a_n be negative for any n? (Think about the first value of n for which a_n is not positive.)

(d) Does this help you with exercise 42?

(e) If $1 + 3rk > -1$ and $h > 0$, which is larger, $C(3h)$ or $C(h)$?

44. If $0 < h < 1$, which is larger, $C(h)$ or $C(-h)$?

The exponential function

For $r = 1$, exercise 40 gives you the values of h shown in table 3.15. You can see that, as h decreases to zero through positive values, $C(h)$ seems

Table 3.15

h		$C(h)$	h	$C(h)$
1.0	(= 1)		−1.0	
0.5	(= 1/2)		−0.5	
0.25	(= 1/4)		−0.25	
0.125	(= 1/8)		−0.125	
	$\vdots$			
	(= 1/1024)			

to be increasing and as h increases to zero through negative values, $C(h)$ seems to be decreasing. As h approaches zero from either side, $C(h)$ seems to be approaching a limit, and it looks as though

$$\lim_{h \to 0} [C(h)] = 2.718\,281\,824\,459\ldots \tag{3.13}$$

as h approaches zero.

Exercises 41–43 lead to rigorous proofs of the statements about the increasing or decreasing behavior of $C(h)$, and exercise 44 explains the comparison between the values of $C(h)$ for positive h with those for negative h.

This makes it plausible that the limit in (3.13) exists and has the approximate value given there. This number is one of the two most important special numbers in mathematics and is denoted by e (after the Swiss mathematician Euler (1707-77), who discovered some of its most important properties). Our tables suggest the 'sandwich'

$$C(h) < e < C(-h) \tag{3.14}$$

for positive h, which enables us to estimate e as closely as we wish.

Applying the result (3.13) to equations (3.10) and (3.11), with $r = 1$, we see that the solution of the differential equation

$$x'(t) = x(t) \tag{3.10a}$$

is

$$x(t) = x(0)e^t, \tag{3.15}$$

Equations (3.11) and (3.12) tell us how to compute this function as accurately as we wish.

More generally, for any given r, the limit

$$\lim_{h \to 0} C(h) = \lim_{h \to 0} (1 + rh)^{1/h} = E(r) \tag{3.16}$$

seems to exist, where $E(r)$ indicates the dependence of the constant E on r. This would imply that the solution of (3.10) is

$$x(t) = x(0) \, E(r)^t. \tag{3.17}$$

We encounter here a special type of function, a constant to a variable power. Such functions are called *exponential functions*. The particular case e^t [$x(0) = 1, r = 1$] is called *the* exponential function. We shall see in a moment that exponential functions behave quite differently from polynomials and the other elementary functions studied in school.

Equations (3.11) and (3.12) tell us one way to compute an exponential function. With a digital computer, it is easier to use the equation

$$x(t + h) = (1 + rh) x(t), \tag{3.18}$$

which you obtain by solving (3.8) for $x(t + h)$. This equation is very simple to program. You take the given value of $x(0)$ as an input, and use equation (3.18) repeatedly for $t = 0, h, 2h, \ldots$, until you reach the value of t in which you are interested.

With either method, roundoff errors gradually accumulate, so one only obtains accurate values of $x(t)$ when t is close to zero. For larger values of t it is more practical to combine these methods with others based on the properties of exponential functions discussed in later sections.

Exercises

45. Find a simple formula for $C(h)/C(-h)$. For $h = 1/n$, $n > 2$, you can apply exercise 43 to obtain a lower estimate for this ratio, and inequality (3.14) gives you an upper estimate. For $h = 2^n$, as n increases $C(-h)$ decreases. Use all this to obtain an upper estimate for $C(-h) - C(h)$, which tells you how good an estimate (3.14) is in fact.

46. Let us use the notation

$$C(h,r) = (1 + r)^{1/h}$$

to show the dependence of $C(h)$ on r as well as h. Obtain a simple formula for $C(h,r)$ in terms of $C(rh,1)$. Use (3.16), and the fact that $E(1) = e$, by definition, to obtain a simple formula for $E(r)$.

47. Obtain a simpler formula for $x(t)$ in (3.17).

Direct study of the differential equation

The technique of approximating a differential equation by a difference equation, which we used in approximating (3.10) by (3.8), can be applied to many other differential equations. This is often the most practical way to calculate the solutions. Thus we can approximate the differential equation

$$x'(t) = 1 + t + x(t)^4 \tag{3.19}$$

by the difference equation

$$\frac{x(t + h) - x(t)}{h} = 1 + t + x(t)^4,$$

which may be put in the form

$$x(t + h) = x(t) + h\,[1 + t + x(t)^4]. \tag{3.20}$$

It is easy to program the computation of $x(t)$, given $x(0)$, for $t = h, 2h, 3h, \ldots$, from equation (3.20). If h is small, then the computed values of $x(t)$ will be close to the values of the exact solution of the differential equation, at least for t in a certain interval.

While there is a simple formula, namely $x(t) = x(0)e^{rt}$, for the solution of (3.10), there is no formula in terms of simple functions for the solution of (3.19). Equation (3.19) can be taken as the definition of a new function. You can tabulate it by using (3.20). If you used this function frequently, it would become as familiar to you as the square root function.

Most differential equations which occur in pure and applied mathematics, like equation (3.19), cannot be solved in terms of simple formulas involving already-known functions. When this does happen, it is a lucky accident. When you have this luck, you should learn to recognize it and know how to take advantage of it, but you should also know how to deal with the usual situation. We shall now discuss how you can learn a great deal about the solution of (3.10), by *direct study* of the equation, *without* solving it. These techniques can be applied to equations like (3.19) whose solutions are more complicated.

First, we remark that the units used in describing most phenomena have no natural meaning and are *merely* matters of human agreement. For example, we could measure time in seconds, days, years or centuries. It is therefore always useful to see whether the problem might be simplified by using other units. If we originally measured time in seconds, and

$$1 \text{ new unit} = k \text{ seconds,}$$

then

$$\theta \text{ new units} = k\theta \text{ seconds,}$$

so that the equation expressing our old measure in terms of the new one is

$$t = k\theta.$$

we then have

$$\frac{dx}{d\theta} = \frac{dx}{dt} \cdot \frac{dt}{d\theta} = k\frac{dx}{dt} = krx,$$

if x is a solution of (3.10). Which choice of k makes this equation as simple as possible?

Clearly the best choice of k is $k = 1/r$, and equation (3.10) is reduced to the special case

$$\frac{dx}{d\theta} = x. \tag{3.10b}$$

Thus it is sufficient to study the case $r = 1$. The equation

$$t = k\theta = (1/r)\theta$$

is equivalent to

$$\theta = rt.$$

If $u(\theta)$ is a solution of (3.10b), then

$$x(t) = u(rt)$$

is a solution of (3.10).

Let $E(\theta)$ be the solution of (3.10b) such that $x(0) = 1$, so that $E(\theta)$ is defined (cf. equation (3.17)) by the equation

$$E'(\theta) = E(\theta), \ E(0) = 1, \tag{3.21}$$

that is, by a differential equation together with an initial condition.

Since $E(0) = 1$, then for θ close to zero, $E(\theta)$ will be close to 1. How soon can $E(\theta)$ reach the value 2? Suppose that T is the first point such that $E(\theta) < 2$ for $0 < \theta < T$, and $E(T) = 2$, as in fig. 3.2. By (3.21) for $0 \leqslant \theta < T$ we have

Fig. 3.2

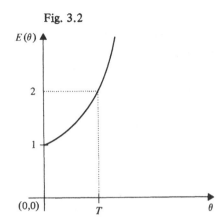

$$E'(\theta) < 2,\tag{3.22}$$

so that

$$[E(\theta) - 2\theta]' = E'(\theta) - 2 < 0.$$

Therefore $E(\theta) - 2\theta$ is decreasing in the interval $0 \leqslant \theta < T$, and

$$E(\theta) - 2\theta < E(0) - 2 \times 0 = 1.$$

This yields

$$E(\theta) < 1 + 2\theta$$

in this interval. Let θ approach T. Then

$$E(T) = 2 \leqslant 1 + 2T$$

so that

$$\frac{1}{2} \leqslant T.$$

Therefore we find that

$$E(\theta) < 2\tag{3.23}$$

for $0 \leqslant \theta < 1/2$, and $E(1/2) \leqslant 2$. This method can be used to get much more information about the solution of equation (3.21).

Exercises

48. Why did we not discuss the case $r = 0$ in (3.10)? Do we need a special theory in this case?

49. In (3.22), 2 is the derivative of what function of θ? How does this explain the next step?

50. If $x(\theta)$ is a solution of (3.10*b*) and $x(0) \geqslant 0$, can $x(\theta)$ be negative for any positive value of θ? Let ϵ be any positive number. Discuss the first point T where $x(T) = -\epsilon$.

51. If $x(0) \geqslant 0$ and $\epsilon > 0$, what can be the first point T where $x(T)$ $= x(0) + \epsilon$? For $0 \leqslant T < 1$, is the inequality $x(T) > x(0)/(1 - T)$ possible? What choice of ϵ would this correspond to?

52. Let $y(\theta) = 3x(\theta)$, which corresponds to a change in the unit for measuring x. What differential equation does $y(\theta)$ satisfy? Generalize to the case $y(\theta) = Cx(\theta)$, where C is any constant.

53. Does the substitution of exercise 52 give a simple result when applied to equation (3.19)?

54. Let $z(s) = x(s + c)$, corresponding to the relation $\theta = s + c$, where c is any constant. What differential equation does $z(s)$ satisfy? Does this substitution give a simple result with equation (3.19)?

55. Prove that if $x(c) \geqslant 0$, then $x(c + s) < x(c)/(1 - s)$ for $0 < s < 1$, where x is any solution of (3.10b) and c is any constant.

56. Prove that if $x(0) \geqslant 0$ and x satisfies (3.10b), then $x(t) \leqslant 4x(0)$ for $0 \leqslant t \leqslant 1$, and $x(t) \leqslant 8x(0)$ for $0 \leqslant t \leqslant 3/2$. What estimate can you obtain for $x(10)$?

57. If x is a solution of (3.10b) and $x(0) = 0$, what is $x(\theta)$ for $0 \leqslant \theta < 1$? (Hint: see exercises 42 and 43.) What about $x(3/2)$ or $x(10)$? (See exercise 56.)

58. If x and X are both solutions of (3.10a) (i.e., $X'(t) = X(t)$), and $u(t) = x(t) + X(t)$, what differential equation does u satisfy?

59. What happens when you apply the method of exercise 58 to equation (3.19)?

From exercise 50 you can see that if x is a solution of (3.10b) and $x(0)$ $\geqslant 0$, then $x(\theta) \geqslant 0$ for all $\theta \geqslant 0$. Then

$$x'(\theta) = x(\theta) \geqslant 0 \text{ for } \theta > 0,$$

that is, x' is always non-negative. Hence $x(\theta)$ is a *non-decreasing* function. In particular, we must have

$$x(0) \leqslant x(\theta) \leqslant x(T) \text{ for } 0 \leqslant \theta \leqslant T. \tag{3.24}$$

Now, the constant $x(0)$ is the derivative of the function $x(0)\theta$, so that

$$[x(\theta) - x(0)\theta]' = x'(\theta) - x(0)$$
$$= x(\theta) - x(0)$$
$$\geqslant 0.$$

Therefore, $x(\theta) - x(0)\theta$ is also non-decreasing, so that

$$x(\theta) - x(0)\theta \geqslant x(0) - x(0) \times 0 = x(0),$$

and

$$x(0)(1 + \theta) \leqslant x(\theta) \text{ for } 0 \leqslant \theta. \tag{3.25}$$

In the same way, we show that

$$x(\theta) \leqslant x(0) + x(T)\theta \text{ for } 0 \leqslant \theta \leqslant T. \tag{3.26}$$

If we apply this inequality for $\theta = T$, we find that if $0 \leqslant T < 1$, then

$$x(T) \leqslant x(0)/(1 - T) \tag{3.27}$$

which we proved in another way in exercise 51. We can continue this process, step by step, by repeating this reasoning.

The left-hand side of (3.25) is the derivative of what function? You can see that

$$g(\theta) = x(\theta) - x(0)\left(\theta + \frac{\theta^2}{2}\right)$$

is a non-decreasing function, so that

$$g(0) \leqslant g(\theta)$$

and

$$x(0)\left(1 + \theta + \frac{\theta^2}{2}\right) \leqslant x(\theta) \text{ for } \theta \geqslant 0. \tag{3.28}$$

If you apply the same reasoning to (3.26), you obtain

$$x(\theta) \leqslant x(0)(1 + \theta) + x(T)\frac{\theta^2}{2} \tag{3.29}$$

for $0 \leqslant t \leqslant T$, Then if $T^2/2 < 1$ (i.e., $0 \leqslant T < \sqrt{2}$), we obtain from (3.29)

$$x(T) \leqslant x(0)(1 + T)\bigg/\left(1 - \frac{T^2}{2}\right). \tag{3.30}$$

Exercises

60. Equations (3.25) and (3.30) give you a sandwich for estimating $x(\theta)$ for $0 \leqslant \theta < \sqrt{2}$. Apply this to $E(\theta)$ ($E(0) = 1$). Estimate the error in using the left-hand side of (3.25) to approximate $E(\theta)$. For which values of θ is the error less than 0.05, 0.005? Estimate the error in using (3.28) for the same values of θ.

61. Improve (3.28) and (3.30) by the same method. The left-hand side of (3.28) is the derivative of what function? What about the right-hand side of (3.29)?

62. Do you see a pattern in (3.24), (3.25), (3.28), and exercise 61? In (3.24), (3.26), (3.29), and exercise 61? Can you predict the result using the method once more? Try it, and check your predictions.

63. Predict the approximations for $E(1) = e$ obtained by applying this method a few more times. Compute to four decimal places. Compare with the method on p. 78. Which is more efficient?

64. (*a*) Obtain some lower estimates for the solution of (3.19) satisfying the initial condition $x(0) = 0$.
    (*b*) Is this solution increasing or decreasing for $t \geqslant 0$?

(c) Find a lower estimate for the positive solution of $x(T) = 1$.

(d) Obtain some upper estimates for $x(t)$ for $0 \leqslant t < T$. How much do the estimates in this exercise differ from each other?

(e) Is $3t + x(t)^{-3}$ increasing or decreasing for $t > 0$? Can a solution $x(t)$, such that $x(0) \geqslant 0$, exist in the interval $0 \leqslant t \leqslant 4/3$? (Hint: compare $x(1)$ with $x(4/3)$.)

The logarithmic function

The differential equation (3.10) can be approached usefully by studying the inverse function, t as a function of x. If $x(0) = 1$, then $x(t)$ is increasing for $t > 0$. Hence for given $x > 1$ there is a unique positive t such that $x(t) = x$ (see fig. 3.3). We can denote this unique value of t by $t(x)$.

Fig. 3.3

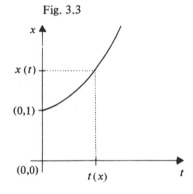

In calculus, one learns the relation between the derivatives of inverse functions:

$$\frac{dt}{dx} = 1 \bigg/ \frac{dx}{dt} \, ,$$

so that equation (3.10) yields

$$\frac{dt}{dx} = \frac{1}{x} \, , \tag{3.31}$$

and the initial condition $x(0) = 1$ yields the initial condition

$$t(1) = 0 \tag{3.32}$$

that is, $t = 0$ when $x = 1$.

The differential equation (3.31) is very special because the unknown function t does not appear on the right-hand side. The derivative of t is simply a known function of x. The problem of finding a function with a given derivative is the basic problem of integration.

It is easy to see that t cannot be a polynomial in x, such as

$$7x^4 - 3x^2 + 2x + 11,$$

since the derivative of a polynomial is again a polynomial. It is a little less easy, but still not difficult, to see that t cannot be a rational function of x, that is, a ratio of two polynomials, such as

$$\frac{7x^4 - 3x^2 + 2x + 11}{3x^5 + 4x^3 + 1}$$

(see the exercises below). So the solution of (3.31) is a new function different from these elementary functions.

Let us call the solution of (3.31) and (3.32) $L(x)$. In calculus, one learns to represent this solution as a definite integral:

$$t = L(x) = \int_1^x \frac{1}{s} \, ds. \tag{3.33}$$

Note that the lower limit in the integral is 1, to fit the initial condition (3.32). Also, we were careful to use different letters for the variable of integration and the upper limit of the integral, so as to avoid confusion.

We can easily obtain a great deal of information about $L(x)$ from (3.33), as well as practical methods for computing it. For example, since for $1 \leqslant s \leqslant x$ the function $1/s$ is decreasing, we see that

$$\frac{1}{x} \leqslant \frac{1}{s} \leqslant 1 \text{ for } 1 \leqslant s \leqslant x.$$

This yields the estimate

$$\frac{x-1}{x} = \int_1^x \frac{1}{x} \, ds \leqslant L(x) \leqslant \int_1^x 1 \, ds = x - 1. \tag{3.34}$$

If x is close to 1, this is a very good sandwich for $L(x)$ and enables us to compute it approximately.

We can obtain better estimates if we first subdivide the interval of integration and then estimate as above. For example, if we divide the interval into three equal parts as in fig. 3.4, each part has length $h = (x-1)/3$, and we obtain

$$L(x) = \int_1^{1+h} \frac{1}{s} \, ds + \int_{1+h}^{1+2h} \frac{1}{s} \, ds + \int_{1+2h}^{1+3h} \frac{1}{s} \, ds,$$

hence

$$h\left(\frac{1}{1+h} + \frac{1}{1+2h} + \frac{1}{1+3h}\right) \leqslant L(x) \leqslant h\left(1 + \frac{1}{1+h} + \frac{1}{1+2h}\right). \tag{3.35}$$

Fig. 3.4

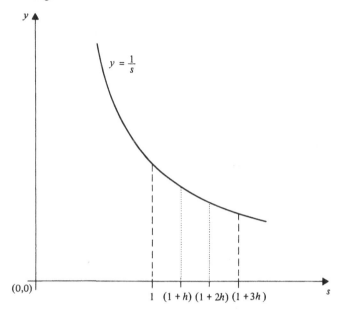

If we use three subintervals, but choose the division points so that they form a geometric progression with common ratio r as in fig. 3.5, then $r = x^{\frac{1}{3}}$, and we obtain

$$(r-1)\frac{1}{r} + (r^2 - r)\frac{1}{r^2} + (r^3 - r^2)\frac{1}{r^3} \leqslant L(x) \leqslant (r-1) + (r^2 - r)\frac{1}{r} + (r^3 - r^2)\frac{1}{r^2}.$$

This gives, by elementary algebra,

$$3x^{-\frac{1}{3}}(x^{\frac{1}{3}} - 1) \leqslant L(x) \leqslant 3(r-1) = 3(x^{\frac{1}{3}} - 1). \qquad (3.36)$$

These estimates give better approximation for $L(x)$ than (3.34). If we sub-divide the interval from 1 to x into even smaller intervals we obtain still better approximations. The same procedures can be used for estimating integrals of any continuous function $g(s)$. While there are some ways to improve this method, basically it is the only general method which works for a quite general integral.

This method is, however, not very efficient. For the special integral $g(s) = 1/s$ there are methods which give high accuracy with much less work, if the interval of integration is not too large. A simple but effective method con-sists of substituting

$$s = 1 + z$$

Fig. 3.5

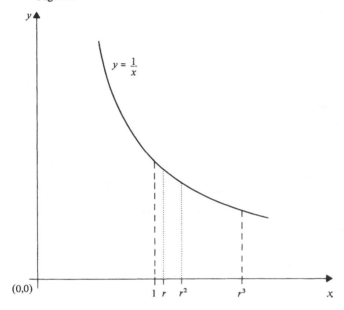

in the above integral and using the abbreviation $u = x - 1$. We obtain

$$ds = dz$$

so that

$$L(x) = \int\limits_{0}^{u} \frac{1}{1+z} \, dz.$$

If we apply the elementary algebraic process of long division

$$
\begin{array}{r}
1 - z + z^2 - z^3 \\
1 + z \overline{)1 } \\
\underline{1 + z} \\
- z \\
\underline{-z - z^2} \\
z^2 \\
z^2 + z^3 \\
\underline{- z^3} \\
- z^3 - z^4 \\
\underline{ z^4}
\end{array}
$$

we obtain

$$\frac{1}{1+z} = 1 - z + z^2 - z^3 + \frac{z^4}{1+z}.$$

This gives us

$$L(x) = u - \frac{u^2}{2} + \frac{u^3}{3} - \frac{u^4}{4} + \int_0^u \frac{z^4}{1+z}\, dz\,.$$

In order to estimate the last integral, we note that for $0 \leqslant z \leqslant u$ we have

$$\frac{z^4}{1+u} \leqslant \frac{z^4}{1+z} \leqslant z^4.$$

Hence we obtain

$$\frac{u^5}{5(1+u)} \leqslant \int_0^u \frac{z^4}{1+z}\, dz \leqslant \frac{u^5}{5},$$

and

$$u - \frac{u^2}{2} + \frac{u^3}{3} - \frac{u^4}{4} + \frac{u^5}{5(1+u)} < L(x) \leqslant u - \frac{u^2}{2} + \frac{u^3}{3} - \frac{u^4}{4} + \frac{u^5}{5}. \qquad (3.37)$$

If $0 \leqslant u \leqslant 0.1$, this sandwich gives us an approximation to $L(x)$ with an error less than

$$\frac{(0.1)^5}{5}\left(1 - \frac{1}{1.01}\right),$$

so that we obtain $L(x)$ correct to six decimal places.

Exercises

65. If $1 \leqslant x \leqslant 1.1$, how big can the difference be between the two sides of the sandwich (3.34)? Give an approximation to $L(1.05)$ and say how accurate it is.

66. Calculate $L(1.064)$ using (3.35) and (3.36). Which equation gives a better approximation? Compare with what you get using (3.37). You may use a hand or desk calculator.

67. Generalize (3.35) and (3.36) by using n subintervals. Estimate the error. If $1 \leqslant x \leqslant 1.1$, how large must n be to obtain a result correct to three decimal places?

68. Obtain formulas like (3.37) involving powers of u up to u^7 and up to u^8. Estimate their accuracy for $1 \leqslant x \leqslant 1.5$ and for $1 \leqslant x \leqslant 2$. Are these formulas useful for $x = 2.1$?

69. (a) Solve the differential equation

$$x'(t) = x^2,$$

with the initial condition $x(0) = 1$. (Hint: consider t as a function of x, and use the method of this section.)

(b) Compute the solution of the difference equation

$$\frac{x(t+h) - x(t)}{h} = x(t)^2,$$

with $x(0) = 1$, for $h = 0.1$. Do you get a good approximation to the solution of (*a*) for $0 \leqslant t \leqslant 0.5$? Is the approximation better if you take $h = 0.05$? For both values of h compute $x(1.1)$ for the solution of the difference equation above. Does this give a good approximation to the solution of (*a*) for $t = 1.1$? Explain.

Difference equations and differential equations

In studying differential equations, as in many branches of mathematics, there are basic facts and side issues. From our point of view, the basic point is to emphasize the real situation, and the misleading approach is the one unfortunately all too common in elementary texts, the cook-book approach, with its recipes and the disproportionate importance it attaches to solutions in closed form.

This is why we studied at such length what we can learn about a differential equation by looking at the corresponding difference equation. After all, this method is not only used in digital computers in order to find approximate solutions, but is also the basis of the proof of existence theorems.

Sometimes the rationale given for using difference equations instead of differential equations for studying the laws of nature is that matter is built up from atoms, so that if x represents the number of members of a population (in its general sense), x is really a discrete variable. A more rigorous treatment would then require that x be considered as a random variable with a probability distribution varying with time. Then equation (3.10), namely $x'(t) = rx(t)$, would be exact for the expected value of x at the time t. Such a probabilistic approach is beyond the scope of this book.

3.5   Models again

We discussed models and their limitations in chapter 2. In this chapter we have examples of several concrete models which have the same mathematical structure. This enhances the value of the mathematical model, since a careful analysis of this one structure will permit us to make predictions for several concrete problems. The whole idea of 'mathematical abstraction' has its roots here.

Another facet of the use of mathematical models is the dual approach: discrete time or continuous time. In our discussion of radioactive decay or of the struggle for life we started by looking at discrete time intervals, then set up difference equations, and ended up with a continuous exponential function, the solution of a differential equation.

The use of hand calculators or of computers has taught us that discrete time models are often easy to treat computationally, and are therefore useful for obtaining numerical results or predictions. On the other hand, a continuous time model, which can be expressed by continuous functions, often makes for simplicity in the description of the phenomena and also in their conceptualization. We express this dual approach graphically in fig. 3.6. The

Fig. 3.6

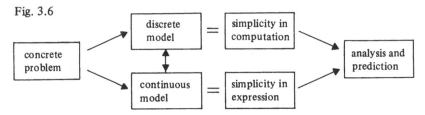

two-way arrow between 'discrete model' and 'continuous model' represents the dilemma of which one is the 'real' model and which the 'approximation'.

The purely pragmatic approach certainly has some logic to back it up. After all, the measuring we do and the information we get about a concrete problem is always of the discrete type. The computing is done on a discrete basis by the computer. Why then bother to look for a continuous model? This way of thinking is expressed in fig. 3.7. To illustrate this approach, we shall set up discrete models for some classical problems in dynamics in a later chapter.

Fig. 3.7

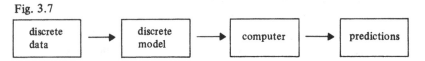

However, our preceding remarks about the conceptual simplicity of a continuous model are certainly as valid as those about the discrete model. Working with continuous models, it will often be much easier for us to predict phenomena, other results, new laws, and in general make progress in our theoretical analysis of phenomena.

Discussion problems

70. Suppose one tried to set up a discrete model of a line as consisting of segments of a certain minimum length. Consider a square whose sides are such 'discrete' lines. Does the diagonal have a definite length? Can a plane be isotropic, that is, can it have the same structure in all directions?

71. Consider the problem of measuring the circumference of a circular wheel. How could you do it practically? How could you describe this process mathematically using a discrete model for a line?

72. If space has a discrete 'crystalline' structure, if light travels in straight lines, and if time is discrete, can the velocity of light be the same in all directions?

4

Descriptive statistics

During the last twenty years statistics has become an integral part of the mathematics curriculum at all levels. In this chapter we propose various real and well-based applications that can be taught from primary school onwards.

We start with a discussion of Zipf's law, an empirical law in linguistics which permits us to show the pupils how to gather and analyze data. Since this is at secondary school level, we illustrate in section 4.2 that the same processes of thought can be explained at an early age. After all, children like to tabulate things and this can be taken advantage of in order to introduce them to meaningful, although very simple, statistics, while at the same time encouraging team-work. In section 4.3 codes and code-breaking methods, using frequency tables, are discussed. The methods we use were practical before the advent of computers, and are still a necessary introduction to the subject. In our experience this subject has always elicited an enthusiastic response. The mathematical knowledge required has been kept to a minimum.

The basic notions of mean, variance, and standard deviation are introduced, starting from a discussion of average height, in section 4.4.

In section 4.5 we emphasize that statistics is a way to describe observations, and that probability is a mathematical model to explain the regularities we find in statistical data. We discuss probabilities of events, probability measure on finite sample spaces, independent events, random variables, and expectation. We discuss, at a level suitable for advanced high school students, the law of large numbers, which shows that the probability model fits the observed data. Finally, in a part suitable for calculus classes, we give a brief treatment of the connection between probability and integration.

4.1   Zipf's law

George Zipf was a professor at Harvard University who loved to count. He counted all sorts of things: pictures in mail-order catalogs, numbers of cities with various populations, and so on. Among the things he counted were words in various texts. He discovered a remarkable law about the frequencies of words, which seems to apply to almost any text in almost any language. You can read about Zipf's law in his book *Human behavior and the principle of least effort* (Addison-Wesley, Reading, Mass., 1949). The law had also been discovered independently by the Swiss linguist Ferdinand de Saussure (1857-1913). For some years Zipf's law remained a purely empirical law. Finally, in 1959, the French mathematician Mandelbrot found a theory which gives a satisfactory explanation of Zipf's law (B. Mandelbrot, 'Statistical macro-linguistics', *Nuovo Cimento, Suppl.*, 13 (1959), 518-20.)

In this section we will show you how Zipf's law was discovered.

Type, token, and rank

The relative frequencies of words affect many areas of speech behavior. The more frequently used words are usually shorter than the rarer words. This can be easily verified, for example, in English, French, German and Spanish. Zipf even observed that the majority of the commonly used words in many languages are monosyllables.

The word 'word' is ambiguous in ordinary usage. The sentence 'the dog bit the man and the man bit the dog back' contains twelve words (one meaning) and the word 'the' occurs four times in it (another meaning). To keep our meaning clear it is best to use two different terms. We shall call the individual words (the first meaning) *token*, and a word in the second meaning a *type*. Thus this sentence contains twelve tokens and only six types. The ratio (here 6/12) of the number of types in a text to the number of tokens is a measure of the diversity of the text. We call it the *type-token ratio*.

Let the most frequently occurring word type be given the rank of one, the second most frequent a rank of two, and so on. Zipf found an interesting relation between rank and frequency.

Exercises

1. Take two pages of any text and arrange the words in order of frequency. Fill in the table:

Word type	Rank	Frequency
	1	
	2	
	.	
	.	
	.	

Let r be the rank, and $f(r)$ the frequency. Plot $f(r)$ against r on graph paper, and compare your results with those of other students in the class. If you have used ordinary graph paper, the result should look somewhat like fig. 4.1, resembling part of a rectangular hyperbola. In

Fig. 4.1

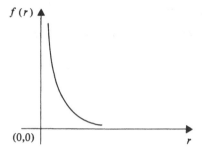

the usual coordinates x and y the equation of such a curve is

$$xy = c, \tag{4.1}$$

where c is a constant.

Using log–log paper

It is easier to discover the relationship between frequency and rank if you plot your data on log–log paper, which was described in chapter 2. If r and f satisfy the equation

$$rf = c,$$

then their logarithms satisfy

$$\log r + \log f = \log c.$$

Thus on log–log paper this will appear as a straight line with slope -1, which therefore makes an angle of $-45°$ with the positive r-axis (fig. 4.2).

Fig. 4.2

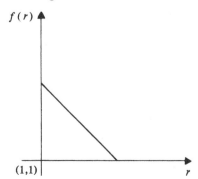

Exercises

2. Add another column to your table in exercise 1 and tabulate there the product rf. Are the numbers in this column approximately constant? Graph f versus r on log-log paper. Is the graph fairly close to a straight line? What angle does it make with the r-axis? Compare your results with those of your classmates.

Figure 4.3 illustrates the rank-frequency distribution of words. Curve A is from James Joyce's *Ulysses* (Bodley Head, London, 1960), which contains about 4 million word tokens and nearly 30 000 word types. Table 4.1 contains the data on which curve A is based. Curve B is based on a count of words in

Fig. 4.3

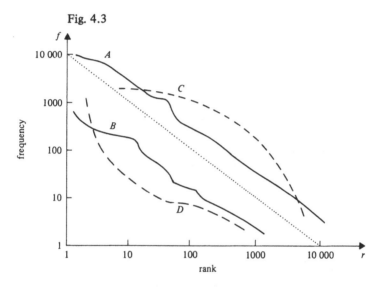

American newspapers that tallied 43 989 word tokens and some 6 000 types.

The importance of the graph is that, despite the different sources of these counts, the two curves A and B are very similar.

Deviation from the straight line indicates abnormality in speech behavior. Line C (which is hypothetical) has a gentler slope and indicates *diversification.* It means a larger vocabulary and a lower probability of occurrence of the most frequent words. Another hypothetical line D has a steeper slope and indicates a simplification, a tendency to use fewer different words. In this case a few words are used again and again, and the probability of occurrence of the most frequent words is greater. Such a trend might occur in children, or speakers with low education or intelligence. (In both cases *slopes* refer to the left ends of lines C and D only.)

Fig. 4.3 shows the likelihood that a certain word will occur *on the average.*

Table 4.1 *Exerpts from a rank-frequency table of James Joyce's 'Ulysses'*

Rank	Frequency	Product (rf)
10	2 653	26 530
20	1 311	26 220
30	926	27 780
40	717	28 680
50	556	27 800
100	265	26 500
200	133	26 600
300	84	25 200
400	62	24 800
500	50	25 000
1 000	26	26 000
2 000	12	24 000
3 000	8	24 000
4 000	6	24 000
5 000	5	25 000
10 000	2	20 000
20 000	1	20 000

If we take into account the likelihood of the word's occurring in a specific context, we do not obtain the curve predicted by $rf = c$. For example, after 'of' the most frequent word is 'the', which occurs ten times as often as any other word in that position. After 'the', however, almost anything can happen, as there is a long list of words that are about equally likely. Thus the curve of words of frequency less than 'of' has a steep slope, while the curve of words of frequency less than 'the' has a gradual slope (fig. 4.4).

Fig. 4.4

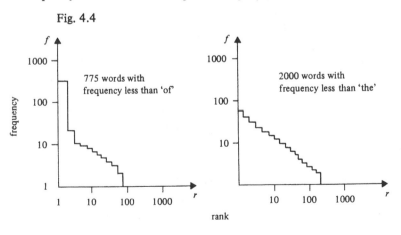

775 words with frequency less than 'of'

2000 words with frequency less than 'the'

Exercises

3. Test the truth of this argument on the curves of selected words in passages of your choice.

Zipf's law states that the distribution of word frequencies, calculated from any ordinary sample of language, always has the same mathematical form. The data from many unrelated languages and from texts covering thousands of years of history have been found to conform to this simple equation. Written as well as spoken language on any subject matter shows this relationship between frequency and rank. Since the equation depends on the number of words occurring with every possible frequency, it is not greatly affected by circumstances that increase the number of words occurring just once; they affect only a single point on the curve.

Some interpretations of Zipf's law

According to Zipf, human behavior in many spheres is founded on the principle of least effort. The organism strives to maintain as low an average level of exertion as possible (e.g., arrangement of typewriter keyboard, preference of certain syllable structures). Zipf claimed that from the speaker's point of view, language would be at its simplest if the speaker had only to utter the same word again and again, or, in other words, if the language consisted of a single word only. With a single-word language, the speaker does not have to go through a selection process when he needs a specific word. Moreover, this one word is so overlearned that it requires very little effort to produce. From the listener's point of view, on the other hand, the language would be most rational and most convenient if every distinctive meaning had its own word. These two tendencies conflict with each other in language. Zipf's standard curve can be viewed as the equilibrium between these two tendencies.

Zipf's law is a perfect example of a law that has been observed but for which no suitable explanation is immediately apparent.

By the very definition of rank, the frequency $f(r)$ must be decreasing as r increases. This alone gives no indication of the precise relationship. Zipf's argument makes the relation $rf(r) = c$ plausible, but is still not a very clear explanation of why the relation should be just this. There exists a good theoretical explanation given by Mandelbrot, but the level of mathematical knowledge it requires does not permit us to discuss it here.

4.2 Statistics of language in elementary schools

One of the natural sources of motivated arithmetic for children is the study of the statistics of their language. As a by-product they also learn some interesting aspects of their language and something about statistical regularities and fluctuations.

The relative frequencies of letters in English are important in the decoding of cryptograms, which is usually fascinating for children. This study might be prefaced by reading *The gold bug* by Edgar Allen Poe, or *The case of the dancing men* by Sir Arthur Conan Doyle.

The class might be divided into sections, each responsible for getting data on two or three letters of the alphabet. Each day for a week, each child could bring in counts of the numbers and frequencies of letters in, say, two paragraphs of any text. It is convenient to divide the text counted into groups of 100 letters and to record for each group the frequency of each letter counted. The counting can be done simply with tally marks.

A pupil's report for one day's count for one letter might look like table 4.2.

Table 4.2. *Number of As in each group of 100 letters*

Number of group	Number of As counted
1	9
2	6
3	3
4	4
5	9
6	8
7	12
8	7

The pupil might also tabulate the cumulative frequencies, as shown in table 4.3. If the child knows about computation with decimals, he could make a

Table 4.3.

Number of letters counted	Number of As (cumulative frequency)
100	9
200	15 (= 9 + 6)
300	18 (= 15 + 3)
400	22
500	31
600	39
700	51
800	58

third column in table 4.3 for relative frequencies (0.09, 15/200 = 0.075, 18/300 = 0.06, etc.).

Relative frequencies can be recorded on scales as in table 4.4. We record there the relative frequencies for the preceding cumulative frequencies. Using graph paper for the table would of course be convenient.

Table 4.4.

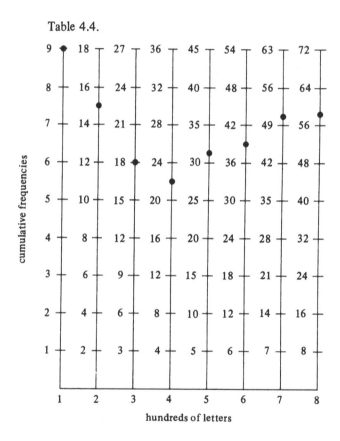

cumulative frequencies

hundreds of letters

The individual daily reports can be combined to form a daily report for each section, and these daily reports can be combined with the results of the previous days to get cumulative section reports. By the end of the week there will be enough data to make a class report on the whole alphabet.

The children can observe many interesting phenomena. For example, in table 4.4 we see both fluctuations and a trend toward about 7.2 As per 100 letters. Will this trend continue as we gather data for 40 hundreds? We also see that one hundred only had 3 As whereas another had 12 As.

In this example the relative frequency fluctuated violently at first, but ultimately settled down to a trend. When we only had a few hundreds, a hundred with only 3 As or as many as 9 As had a big effect on the total. After eight hundreds, *one* group with only a few As or many As will not make such a big difference. The same phenomenon can be observed by comparing batting averages at the beginning of a baseball season with those near the end.

Of course, it might be an interesting task to compose as long a text as possible without a single A, or to write a text with as many As per hundred letters as one can. The statistics we are gathering, however, concern ordinary English as it occurs naturally. It is important to sample a variety of texts to be sure that the data really represent typical usage.

Other interesting questions to investigate are:

> the frequencies of words;
> the frequency of, say, 3-letter words;
> the lengths of sentences;
> the frequency of 3-syllable words.

Some literary critics have tried to compare styles of different authors by such statistics as lengths of sentences or frequencies of words. For example, they have tried to settle in this way arguments about whether certain parts of Shakespeare's plays were really written by Shakespeare.

People who want to develop computers to translate, say, from Russian into English, have tried to analyze English grammar in such a way that it can be programmed into a computer. One approach which they have tried is to set up rules for which word can follow a sequence of some given number of words. For example, they think that if every sequence of five words makes good sense, then there is a good chance that the whole text will be fairly good English. The linguist N. Chomsky has published a proof that no such set of rules can give a complete analysis of English grammar; see, for example, his book *The logical structure of linguistic theory* (Plenum Press, New York, 1975).

It is an amusing task to test this type of hypothesis. One hands a roll of paper tape to the first pupil. He writes a 5-word phrase which makes sense, cuts off the first word, and passes the roll to the next pupil. He adds one word after the other four so that the five make sense, cuts off the first, and passes the roll to the next child. They continue in this way until everyone has written a word. One then writes the whole text on the blackboard, each child in order contributing the word he has kept. How much sense does the whole make? Does it approximate correct English?

The class might compare the results with those of using 4- or 6-word sequences. Are the differences very noticeable?

4.3 Cryptography I

Codes
We all have to send messages sometimes that we want to be read by the addressee only. One of the possible ways to attain this goal is to use a code. The simplest codes are obtained by replacing each letter of the message by another letter. We might, for instance, replace each letter by the letter following it two places further away in the alphabet.

To do the enciphering (writing in code) and the deciphering (returning to standard language) easily, we write the usual alphabet and the code alphabet one below the other:

Plain   A B C D E . . .
Cipher C D E F G . . .

Suppose the message is

I WILL MEET YOU AT MIDNIGHT.

In code it will look like this:

K YKNN OGGV . . .

Finish the coded message.

With this arrangement, enciphering consists of replacing each letter by the one below, and deciphering goes exactly the opposite way. Which letters in our code correspond to Y and Z? Of course, instead of shifting by two places, we can shift by any amount up to 26, as long as both the sender and the receiver are using the same number. Any alphabet obtained in this way is called a 'direct standard alphabet'.

Exercises
4. Decipher the following message:

   D ORW RI JRRG LW ZLOO GR BRX.

5. Devise a general method for deciphering any message which *has been encoded* in a direct standard alphabet. Did you find a good method?

The usual way to decipher is to choose one word of the message and just continue the alphabet under every letter until a meaningful word comes up; if the message were PX PXGM AHFX, we would proceed as follows:

```
PX  PXGM  AHFX
    QYHN
    RZ I O
    S A J P
    TB KQ
    UC LR
    VDMS
    WENT
```

How do you continue? Why does this method work?

Scrambled alphabets

It is clear by now that encoding a message in a direct standard alphabet does not ensure any secrecy. It is too easy to break the code. So let us try
a completely scrambled code. We just decide on a random pairing between the
plain alphabet and the cipher alphabet. As we did before, we write one below
the other:

Plain   A B C D E F G H I J K L M N O P Q R S T U V W X Y Z
Cipher   P D I K W V Z O YBA X C F L H Q E N UG J RM S T

The message

VICTORY IS AT HAND

will be encoded as

JYIULES ...

Finish the coded message.

Trying to decode a message, if you know the code, is best done by first
rewriting the two alphabets in the opposite order, the cipher on top and the
plain below:

Cipher   A B C D E F G H I J K L M N O P Q R S T U V W X Y Z
Plain   K J M B R N U P C V D O X S H AQW YZT F EL I G

Now again we can easily replace each letter in a message by the letter below,
and decode speedily.

Completely scrambled alphabets are not practical, since they require either
memorizing the complete pairing (instead of just a number for the direct
standard alphabet) or keeping a written copy of the code – not very good for
secrecy.

In general, this problem is solved by using a 'keyword', for instance,
POLITICS, writing first the keyword without repeated letters and then the
remaining letters of the alphabet on successive lines under the keyword:

```
P O L I T C S
A B D E F G H
J K M N Q R U
V W X Y Z
```

Then the letters are used as the cipher alphabet, reading them off column after column, like this:

Plain   A B C D E F G H I J K L M N O P Q R S T U V W X Y Z
Cipher   P A J V O B K W L D M X I E N Y T F Q Z C G R S H U

Exercises

6. Encode the message

WE WILL ARRIVE AT NOON

in a scrambled alphabet with keyword CONSTITUTION.

7. Try to decode the following message, written in a scrambled alphabet:

QNFFH RO VLV ENZ BNNX HNC.

Is there a keyword?

We bet that you only succeeded in solving exercise 7 if you have enough insight into the psychology of book-writing mathematicians. It is in fact extremely difficult to decode such a short message, although some tentative conclusions could be drawn from looking at the message and knowing that it is in English, reminding ourselves of particularities of this language.

For instance, consider vowels. There are few English words without vowels. Indeed, looking at the distribution of vowels in any text, for example,

indeed looking at the distribution of vowels

we notice that we rarely have more than two consonants between vowels, that repeated letters in the middle of a word are often vowels, and so on.

Therefore in exercise 7 we would guess that N stands for one of the vowels. Since in BNNX, N appears twice, we reasonably guess

Let us insist here that these are speculative deductions only. The word BNNX might of course be a person's name, but otherwise do you know many English words of four letters with a repeated consonant in the middle? Make a list.

This partial analysis shows that decoding is *guessing*, but educated guessing. To try to do it right we have to know more about our language.

Exercises

8. Take a 1000-letter sample of a plain text (e.g., a newspaper) and make a count of the number of times each letter appears:

A.. E.. I.. M.. Q.. U.. Y..
B.. F.. J.. N.. R.. V.. Z..
C.. G.. K.. O.. S.. W..
D.. H.. L.. P.. T.. X..

9. We call the number of occurrences of a letter its *frequency*. Draw up a bar graph of the frequencies, as in fig. 4.5. Here we have supposed that A has a frequency 7 and S frequency 6. Fill in the blanks. This graph is called the *monographic frequency distribution.*

Fig. 4.5

A B C D E F G H I J K L M N O P Q R S T U V W X Y Z

Frequency distribution

Of course the relative frequencies of the different letters may vary for different texts, but in the main the differences are small. Fig. 4.6 shows a typical frequency distribution. There is a definite pattern of highs and lows;

Fig. 4.6

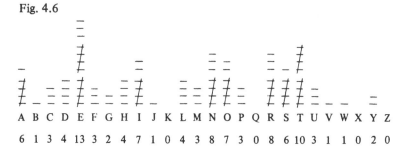

A B C D E F G H I J K L M N O P Q R S T U V W X Y Z

6 1 3 4 13 3 2 4 7 1 0 4 3 8 7 3 0 8 6 10 3 1 1 0 2 0

A, E, I, O are high, as are the consonants T, N, R, S, while on the other hand B, J, K, Q, V, W, X, Z are very low.

Studies of the same kind have been made for digraphs and trigraphs (groups of two or three letters), word beginnings and endings, and so on. The most frequent initial and final letters, for a count made on 1000 words, are shown in table 4.5. The ten most frequent digraphs and trigraphs, for a count of 50 000 characters, are given in table 4.6.

Table 4.5

Initial letter		Final letter	
T	1634	E	2078
A	1126	S	1298
S	758	D	1031
O	735	N	995
I	576	T	992
C	573	R	566

Table 4.6

Digraph		Trigraph	
TH	1351	THE	1073
HE	1283	AND	302
IN	969	TIO	240
ER	898	ATI	180
RE	800	FOR	177
ON	770	THA	159
AN	760	TER	145
EN	643	RES	137
AT	637	ERE	132
ES	573	CON	129

Let us show how to use all this information on an example:

YAZ EDDKWHN CBYAK YRDZCKWYH P
FCS BDDH RYSKRYHDO UZYE
UZWOCI KY SAHOCI HDMK TDDG CHO
TWXX BD FDXO CS ASACX WH KFD
KFWZO UXYYZ QYHUDZDHQD ZYYE.

We start with a frequency count:

5 3 8 15 3 4 1 10 2 0 8 0 1 1 6 1 2 3 5 2 4 0 6 5 13 8
A B C D E F G H I J K L M N O P Q R S T U V W X Y Z

There are 116 letters altogether. The most frequent are:

D	15
Y	13
H	10
Z, K, C	8
O, W	6
S, X, A	5

Now we look at the distribution of the letters in the message. We notice that in this message D and Y appear in the places usually occupied by vowels, for instance, in BDDH, HDMK, CBYAK, ZYYE, and that H, Z, K and O look like consonants. (Why?)

Since both Y and D appear in groups of two this suggests that

From examining fig. 4.6 we can guess that

The distribution of repeated initial and final letters in the coded message is:

Initial   U, K (3)
Final   O (4); D, H (3)

Taking table 4.5 of initial and final letters into account, we would choose pairings for D between E and O and for K, O and H among T, N, R and S as follows:

> for K, a frequent initial: T;
> for H, a frequent final: S or N;
> for D, a frequent final: E and not O;
> for O, a frequent final: N, D, or S.

Tentatively, therefore, we will try:

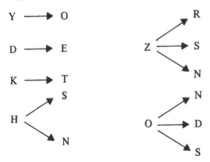

The word KY in the message becomes TO, which looks good. The message now looks like this:

$$\begin{array}{llll} \text{R} & & \text{R} & \\ \text{O-S} & \text{-EET}_{N}^{S}\text{-} & \text{--O-T} & \text{O-ES-T-O}_{N}^{S} \quad - \\ \text{N} & & \text{N} & \end{array}$$

$$\cdots \quad \text{-EE}_{N}^{S} \quad \text{-O-T-O}_{N}^{\overset{N}{S}}\text{ED} \quad \text{etc.}$$

The fourth word suggests to try ON at the end, so let us try H → N. Now the pairings we use are:

$$\begin{array}{ll}
Y \longrightarrow O & O \diagup^{D}_{\diagdown S} \\
D \longrightarrow E & \\
K \longrightarrow T & Z \diagup^{R}_{\diagdown S} \\
H \longrightarrow N &
\end{array}$$

We have a word HDMK in the message, which becomes NE-T. This suggests M → X, quite in accord with its frequency. So the passage in the message around this word looks like

$$\text{TO} \quad \text{--N}_{D}^{S}\text{--} \quad \text{NEXT} \quad \text{-EE-} \quad \text{-N}_{D}^{S}$$

This last word would make sense as AND; therefore we try C → A and O → D. We get

$$\begin{array}{llllll} \text{TO} & \text{--NDA-} & \text{NEXT} & \text{-EE-} & \text{AND} & \\ \text{----} & \text{-E} & \text{-E-D} & \text{A-} & \text{---A-} & \text{-N} \quad \text{etc.} \end{array}$$

The last word should be AN, ON, or IN, but in our pairings O and A are out, so we try W → I, which gives

$$\begin{array}{llllll} \text{TO} & \text{--NDA-} & \text{NEXT} & \text{-EE-} & \text{AND} & \\ \text{-I--} & \text{-E} & \text{-E-D} & \text{A-} & \text{---A-} & \text{IN} \quad \text{etc.} \end{array}$$

The third from last word must be AT or AN or AS; it can only be AS, so S → S. Where are we now?

$$\begin{array}{lllll}
Y \to O & K \to T & M \to X & O \to D & S \to S \\
D \to E & H \to N & C \to A & W \to I & Z \to R
\end{array}$$

Thus we get

$$\begin{array}{lllll} \text{TO} & \text{--NDA-} & \text{NEXT} & \text{-EE-} & \text{AND} \\ \text{-I--} & \text{-E} & \text{-E-D} & \text{AS} & \text{-S-A-} \quad \text{IN} \quad \text{THE} \\ \text{THIRD} & \text{--OOR} & \text{etc.} & & \end{array}$$

including a new deduction that F → H. From here on it becomes a very simple exercise to complete the message. Can you find the keyword? Which letter corresponds to the code letter P?

Exercises

10. Decode the following message and find the keyword:

SO WPYO ZWBZ JPA AFUOGQZBFU ZWO YGOKOULFM
KWBYZOG PF KGJYZPMGBYWJ BDZOG BXX LZ LQ FPZ
ZPP ULDDLKAXZ SOXX LD JPA QAKKOOUOD LF
UOKPULFM ZWLQ IOQQBMO JPA KBFFPZ CO UPLFM
ZPP CBUXJ.

11. This was too easy. To make it more difficult, the encoder would not
leave the message in groups of letters which correspond to words,
but would transmit his message in groups of five letters. Can you
decode this?

BDZAH   MQLGM   ZQILR   LZZVA   OERAE   OQDOQ
UAHGB   HZJAR   ZAHUB   JSAVA   KLUAD   ZQUBJ
ZVBZB   ZQEHM   DZVAC   ARZVA   XBHZO   AGZRV
QXRSQ   EWUCA   LGQHU   AHRQS   AWADZ   QEHKB
HBZZV   AVQZA   WBGUZ   QQPZV   ARECS   BJRZH
BLMVZ   ZQOBK   JR.

It certainly was more difficult. Did you find the relevant keyword?

12. Let us use a completely different way of encoding a message. The
successive stages in the encoding are:

(*a*) Replace each letter by its numerical value (A = 1, B = 2,...,
Z = 26).

(*b*) Add or subtract a constant number from each word (read as
a number).

(*c*) Group all words together in one long string of digits, separating
two words by the string 0110.

For instance, if our constant is +130 071, the message 'we will visit
you at noon' becomes:

   stage (*a*): 235 2391212 22919920 251521 120 14151514
   stage (*b*): 130306 2521283 23049991 381592 130191
14281585
   stage (*c*): 1303060110252128301102304999101103815920110130191011014281585.

Is this a good code? Explain, and give examples.

4.4 Means and medians

How to characterize average height

Let S be the set of all living American males. Let $h(x)$ be the height
of male x, measured in centimeters. Then h assigns a definite number to each

member of S. You can determine $h(x)$ by measuring the individual x, if you catch him! How much work is needed to obtain the value of $h(x)$ for each x? Even if you had this information it would not be particularly useful. It is hard for the unaided human mind to grasp the significance of some 90 000 000 numbers.

Even if you could obtain the height of each x, you would want to summarize the information in terms of a few numbers whose meanings are easy to understand, such as:

the average height;
the minimum height;
the maximum height;
a measure of the variability of $h(x)$, for instance the percentage
with height $\leqslant q$, for q = 100, 110, 120, 130, ..., 200.

We would also want methods of estimating such numbers by measuring only a sample of the whole population S. We would like to know how large a sample to take to obtain an estimate with a certain reliability. We would like some way of testing a sample to tell whether it is representative or not.

Let us examine the first question. What does 'average height' mean? There are many kinds of averages, and each is the most useful for a certain purpose. For instance, suppose we consider the employees of a corporation with the following annual salaries:

5 at $4000,
4 at $10 000
1 at $200 000.

Half the employees have salaries less than $7000, half have larger salaries. The payroll of the corporation works out to $26 000 per man, yet 90% of the employees make less than this. For which purpose is each of these two averages useful?

Let us take a simpler example. Suppose we have five numbers, 5, 7, 4, 2, and 8. What single number x best represents this set of numbers? What should we mean by 'best'?

The errors in using x to represent this set are, respectively, $5 - x$, $7 - x$, $4 - x$, $2 - x$, and $8 - x$. If $x = 6$, then two errors are positive and three are negative. There is a certain choice of x for which there are as many positive as negative errors (one will be neither positive nor negative). Which value of x is it? This is called the *median* of the set: there are just as many numbers in the set less than the median as there are greater than the median. If the set has six members, then any number between the third and fourth, in order of magnitude, has the above property of the median. It is customary then to define the median as halfway between the middle two numbers. Thus the median of the set $\{5,7,4,2,8,6\}$ is 5.5.

The median is a useful average when we wish to avoid giving undue influence to a few exceptional extreme values. Also, in some tabulations of data the numbers are not all given. For income tax purposes a corporation might list its salaries like this:

under $4000	3 employees
$4000	2 employees
$10 000	4 employees
$200 000	1 employee

It would still be easy to calculate the median from this table.

Alternatively we might choose x so that the total error is zero:

$$(5 - x) + (7 - x) + (4 - x) + (2 - x) + (8 - x) = 0.$$

If we solve for x, we obtain $x = 5.2$. This average is called the *arithmetic mean* of the given set. It is easy to compute when all the numbers are known, and algebraic manipulations with the arithmetic mean are convenient. This average is so commonly used that the word 'mean', unless something else is specified, is usually interpreted as the arithmetic mean.

Usually we do not mind whether the errors are positive or negative; it is their numerical values that are important. There are two common ways to measure the size of an error independent of its sign.

One way is to use the *absolute value*. We define $|x|$, the absolute value of x, as follows:

$$\text{if } x \geqslant 0, \text{ then } |x| = |-x| = x.$$

Thus, we have

$$|+3| = |-3| = 3, |0| = 0, \text{ etc.}$$

One natural approach to defining an average is to try to minimize the sum of the absolute values of the errors. In our example, we would look for the x for which

$$|5 - x| + |7 - x| + |4 - x| + |2 - x| + |8 - x|$$

is a minimum.

Instead of using absolute values, one could also measure the size of an error by its square, since $(-a)^2 = a^2$. This leads us to try to minimize the sum of the squares of the errors:

$$(5 - x)^2 + (7 - x)^2 + (4 - x)^2 + (2 - x)^2 + (8 - x)^2.$$

This leads to simpler algebra than if we minimize the sum of the absolute values of the errors.

In the exercises below we need to count the number of elements of some sets. We will denote by $N(A)$ the number of elements of the set A.

Exercises

13. Make a graph of the relation
$$y = |5 - x| + |7 - x| + |4 - x| + |2 - x| + |8 - x|$$
as follows:
(a) If $2 < x < 4$, which of the errors are positive and which negative? Is $|2 - x| = 2 - x$ or $|2 - x| = x - 2$? For x in this interval, express y without absolute value signs. It is then easy to make a graph for x in this interval. When x increases in this interval, does y increase or decrease? For which value of x in this interval is y a minimum?
(b) Carry out a similar analysis for each of the intervals $4 < x < 5$, $5 < x < 7$, $7 < x < 8$, and the half-lines $x < 2$ and $8 < x$.
(c) For which value of x is y a minimum? Do you recognize this value of x?

14. Repeat the previous exercise for the equation
$$y = |5 - x| + |7 - x| + |4 - x| + |2 - x| + |8 - x| + |6 - x|.$$

15. Make a graph of the relation
$$y = (5 - x)^2 + (7 - x)^2 + (4 - x)^2 + (2 - x)^2 + (8 - x)^2$$
as follows:
(a) Express y in the form
$$y = ax^2 + bx + c,$$
where a, b, and c are constants.
(b) Make a table of values

x	0	1	-1	2	-2	3	...
y							

and draw a graph. Can you estimate the value of x for which y is a minimum? Can you recognize the curve?
(c) Find constants h and k such that
$$y = a(x - h)^2 + k$$
for all x. (Hint: $a(x - h)^2 + k = ax^2 - 2ahx + ah^2 + k$.) Comparing with the expression for y obtained in (a) yields $b = ($ $)$, $c = ($ $)$. Fill in the missing values. Solve for h and k.
(d) What is the minimum of $(x - h)^2$? What is the minimum of y? For which value of x is it attained? Do you recognize this value of x?

16. Let $A = \{2,3,10\}$ and $B = \{30,31,32,33,34\}$. Compute the means of the sets A, B, and $A \cup B$. If you already do know
    (a) the means of A and B, and
    (b) $N(A)$ and $N(B)$,
    is there a short cut for computing the mean of $A \cup B$?

17. If B is as in exercise 16 and C is the set of numbers obtained by subtracting 30 from each member of B, what is the relation between the means of the sets B and C?

18. If $D = \{3.0, 3.1, 3.2, 3.3, 3.4\}$, and B is obtained by multiplying each member of D by 10, what is the relation between the means of the sets B and D? (Note: this is the same set B as in exercise 16.)

19. Let a and b be any positive numbers, and let m be the mean of the set $\{a,b\}$. Letting $I(n)$ denote the information function of chapter 2, can $I(m)$ be smaller than the mean of $\{I(a), I(b)\}$? (Hint: what is the sign of $(a + b)^2/4 - ab$?)

20. In exercise 19, which is larger, m^2 or the mean of $\{a^2, b^2\}$?

21. In exercise 19, which is larger, m^3 or the mean of $\{a^3, b^3\}$?

Calculating the mean

We can summarize the process of calculating the mean height of the population S by the formula

$$\text{mean of } h = \frac{1}{N(S)} \sum_{x \in S} h(x)$$

The Greek letter Σ (capital sigma) stands for 'sum'. The formula says to take the value of $h(x)$ for each x in S, sum the results and divide by the number of members in S. The most common notations for the mean of h are $\langle h \rangle$, $\bar{h}$, and $E(h)$. The first is usually used by physicists, the second by statisticians, and the third by probability theorists. Sometimes we indicate the population S over which the mean is taken by $\langle h \rangle_S$.

Of course, we can use the above notation for means of all sorts of quantities.

Exercises

22. Let S be the set of integers from 1 to n, and let $h(x) = 2x - 1$ for all x in S. Compute $\langle h \rangle$ for $n = 2,3,4,5,6$. Can you guess at a formula for any n?

23. Let S be as in exercise 22, and let $h(x) = x^3 - (x - 1)^3$ for all x in S. Compute $\langle h \rangle$ for $n = 2, \ldots, 6$. What is the general rule?

24. Let S be as in exercise 22 and let $h_k(x) = x^k$. Compute $\langle h_k \rangle$ for $n = 10$ and $k = 0,1,2,3$. Also compute $\langle 3h_2 - 3h_1 + h_0 \rangle$. Compare

this result with $3\langle h_2 \rangle - 3\langle h_1 \rangle + \langle h_0 \rangle$. Also compare this result with that of exercise 23. Explain.

25. Using the ideas suggested by exercises 23 and 24, find a formula for $\langle h_3 \rangle$ for any n.

26. (a) Expand $[h(x) - t]^2 = h(x)^2 + (\quad)th(x) + (\quad)$. Fill in the missing values.

(b) If t is any number, find a, b, and c such that

$$\sum_{x \in S} [h(x) - t]^2 = at^2 + bt + c.$$

Find A, B, and C such that

$$\langle (h - t)^2 \rangle = At^2 + Bt + C.$$

(c) For which value of t is $\langle (h - t)^2 \rangle$ a minimum? What is the minimum value of this expression?

The minimum of

$$\langle (h - t)^2 \rangle$$

is a measure of the dispersion, or variability, of h. It is called the *variance* of h. denoted by var(h). The square root of the variance,

$$\sigma(h) = + [\text{var}(h)]^{\frac{1}{2}},$$

is called the *standard deviation* and is also a measure of the dispersion of h.

27. (a) Compute $\langle 3h \rangle - 3\langle h \rangle$, $\langle h + 5 \rangle - \langle h \rangle$.

(b) Compute var($3h$)/var(h), var($h + 5$)/var(h).

(c) Compute $\sigma(3h)/\sigma(h)$, $\sigma(h + 5)/\sigma(h)$.

28. (a) If $S = \{1, 2, 3, 4, 5\}$ and $h(1) = h(2) = h(3) = 0, h(4) = c$, and $h(5) = -c$, compute $\langle h \rangle$, var(h) and $\sigma(h)$.

(b) If $S = \{1, 2, 3, 4, 5\}$ and $h(1) = -3, h(2) = 1, h(3) = 2, h(4) = c$, $h(5) = -c$, compute $\langle h \rangle$, var(h), and $\sigma(h)$.

29. If $\langle h \rangle = 5$, what is the smallest possible value of var(h)? When is this minimum attained? What is the smallest possible value of $\langle h^2 \rangle$?

30. (a) Suppose that $\langle h \rangle = 100, N(S) = 20$, and $h(x) \geqslant 0$ for all x in S. Let A be the set of x in S for which $h(x) \geqslant 900$. What is the greatest possible value for $N(A)$?

(b) Suppose that $N(S) = 20$ and var(h) = 100. Let B be the set of x in S such that $[h(x) - \langle h \rangle]^2 \geqslant 900$. What is the greatest possible value for $N(B)$?

(c) Suppose var(h) = 100 but $N(S) = n$ is unknown. Let B be the set of x in S such that $|h(x) - \langle h \rangle| \geqslant 30$. What is the greatest possible value for the ratio $N(B)/n$?

Properties of the mean

An advantage of the arithmetic mean is that it has simple algebraic properties. Some of these are suggested by the above exercises.

Let us summarize briefly the general meaning of the mean. We have a set S with a finite number of members. A *function* h on S is a way of assigning a definite number $h(x)$ to each x of S. The operation of finding the mean is a way of assigning a definite number $\langle h \rangle = E(h)$ to every function on S. The following rules describe the process of calculating $\langle h \rangle$:

(*a*) Find $h(x)$ for each x in S.
(*b*) Sum the numbers $h(x)$ for all x in S.
(*c*) Divide by $N(S)$.

If f and h are functions on S, we can operate on them to produce new functions in several ways, such as:

$f + h$ is the function whose value for each x in S is $f(x) + h(x)$,
or briefly

$$(f + h)(x) = f(x) + h(x).$$

Similarly we can define for each x in S:

$$(f - h)(x) = f(x) - h(x),$$
$$(fh)(x) = f(x) \cdot h(x),$$
$$(f/h)(x) = f(x)/h(x) \quad (\text{if } h(x) \neq 0).$$

If $f(x)$ has the same value for all x in S, we say that f is a constant. When there is no danger of ambiguity, we can use the symbol '3' for the function whose constant value is 3, and so on.

A *characteristic* function X_A of a subset A of S is simply a function whose values are either 0 or 1. The value 0 is given to X_A for any element of S which is not in A. For $x \in A$ we have $X_A(x) = 1$.

The variance of h can be described simply by the formula

$$\text{var}(h) = \langle (h - \langle h \rangle)^2 \rangle$$

and the standard deviation by the formula

$$\sigma(h) = + [\text{var}(h)]^{\frac{1}{2}}$$

(see exercise 26).

Exercises

31. If A is a subset of S, compute $\langle X_A \rangle$.
32. Compute the following:
    (*a*) $\langle 3 \rangle$ and, more generally, $\langle c \rangle$ if c denotes a constant function.
    (*b*) $\text{var}(c)$ if c is a constant.
    (*c*) $\langle cf \rangle - c\langle f \rangle$ if c is a constant.
    (*d*) $\text{var}(cf)/\text{var}(f)$ and $\sigma(cf)/\sigma(f)$ if c is a constant.

(e) $\langle f + c \rangle - \langle f \rangle$ if c is a constant.

(f) $\text{var}(f + c) - \text{var}(f)$ if c is a constant

(g) $\langle f + h \rangle - \langle f \rangle - \langle h \rangle$.

33. (a) Find a, b, and c such that $\langle (f - th)^2 \rangle = at^2 + bt + c$ for all numbers t.

  (b) If f and h are given functions on S, find the minimum of $\langle (f - th)^2 \rangle$. For which value of t is the minimum attained?

34. (a) Compute $\text{var}(f + h) - \text{var}(f) - \text{var}(h)$.

  (b) Compute
  $$[\langle (f + h)^2 \rangle - \langle (f - h)^2 \rangle]/4.$$

  (c) Compute
  $$[\langle (f + h)^2 \rangle + \langle (f - h)^2 \rangle]/[\langle f^2 \rangle + \langle h^2 \rangle].$$

  (d) Give a formula for $\text{var}(h)$ in terms of $\langle h \rangle$ and $\langle h^2 \rangle$.

35. (a) Can the minimum in exercise 33(b) be negative?

  (b) Given the values of $\langle f^2 \rangle$ and $\langle h^2 \rangle$, what is the greatest possible value for $\langle fh \rangle^2$?

  (c) Given the values of $\langle f^2 \rangle$ and $\langle h^2 \rangle$, what is the greatest possible value for $\langle (f + h)^2 \rangle$?

  (d) Given the values of $\sigma(f)$ and $\sigma(h)$, what is the greatest possible value for $\sigma(f + h)$?

4.5 From statistics to probability

Observations and theory

If you toss a die, it may come up 1, 2, 3, 4, 5, or 6. You cannot in general, predict the result. If the die is honest, and you toss it a large number of times, you will notice some regularities, even though the results of the individual tosses can be any of the above six outcomes.

One experimenter tossed a die 200 times. In some sequences of 10 tosses, 1 or 2 came up seven times, and in others he never had 1 or 2. The relative frequency of the event $X < 3$, where X is the number which comes up, varied from 0 to 0.7. Then he tossed the die 1000 times. Now the relative frequency of the event $X < 3$ in sequences of 50 tosses ranged between 0.20 and 0.48. When he tossed the die 5000 times, the relative frequency of the event $X < 3$ in sequences of 250 tosses ranged only from 0.276 to 0.372.

A similar phenomenon occurs with many events whose outcome seems to depend on chance. For any such event A, the relative frequency $F(A)$ in N independent trials seems to tend to a definite number as N grows larger. This suggests a certain *lawfulness in mass events*, that is, in the outcome of a large number of trials, even though we cannot predict the outcome of any one trial.

This leads one to the idea of developing a theory to describe random outcomes, by means of which we might predict mass events. Thus we might predict that on tossing the die 100 000 times, the number of occurrences of $X < 3$ will be between 27 000 and 38 000 (we would expect about 33 000, so this prediction seems quite safe).

We shall give a brief introduction to the theory of probability. It started about 325 years ago with the work of Pascal and Fermat on games of chance. Nowadays the theory is fundamental to two large industries – insurance and gambling. It is a basic tool in physics, biology, and economics, in the design of agricultural experiments, in business decision-making, and in quality control in industry.

The mathematical model

Suppose we perform N trials of tossing the die. On any one trial the number X which comes up will be one of the six numbers 1, 2, 3, 4, 5, or 6. We call the events.

$$X = 1, X = 2, \ldots, X = 6$$

the *simple events* in this experiment. The set S of all simple events is called the *sample space*. (The sample space is, to be precise, the collection of all simple events. Since we plan to consider only experiments with a finite number of outcomes, we consider the sample space as a set.)

The events

$$X < 3$$

or

$$X \text{ is even}$$

are *compound events* (or *just* events); they are combinations of simple events. For example, $X < 3$ is equivalent to $X = 1$ or $X = 2$, X is even is equivalent to $X = 2$ or $X = 4$ or $X = 6$, and these may be associated with the sets $\{1,2\}$ and $\{2,4,6\}$ respectively.

The frequency $f(A)$ of the event A is the number of times A occurs in our N trials, and the relative frequency $F(A)$ is the ratio:

$$F(A) = f(A)/N.$$

Since

$$0 \leqslant f(A) \leqslant N,$$

then we always have

$$0 \leqslant F(A) \leqslant 1.$$

It is certain that some simple event will occur, so that

$$F(S) = 1.$$

An impossible event, such as $X < 1$, is associated with the null set (symbolized by ϕ), and

$$F(\phi) = 0$$

since it never occurs.

The event

$$X < 3 \text{ and } X > 5$$

is impossible as the sets $\{1,2\}$ and $\{6\}$ have no common members, so that

$$f(X < 3 \text{ or } X > 5) = f(X < 3) + f(X > 5)$$

and

$$F(X < 3 \text{ or } X > 5) = F(X < 3) + F(X > 5).$$

In fact,

$$F(X < 3) = F(X = 1) + F(X = 2)$$

and

$$F(X > 5) = F(X = 6),$$

while

$$F(X < 3 \text{ or } X > 5) = F(X = 1) + F(X = 2) + F(X = 6).$$

In general, if A and B are incompatible events, that is, $\{A \text{ and } B\} = \phi$, then

$$F(A \text{ or } B) = F(A) + F(B).$$

We conjecture that as $N \to \infty$, $F(A)$ approaches a limit, which we shall denote by $P(A)$. We call $P(A)$ the *probability* of the event A. Then $P(A)$ should have the properties

$$0 \leqslant P(A) \leqslant 1,$$

$$P(S) = 1, P(\phi) = 0,$$

and

$$P(A \text{ or } B) = P(A) + P(B) \text{ if } \{A \text{ and } B\} = \phi.$$

We shall call any function P, which assigns to each event A a number $P(A)$ and which satisfies these conditions, a *probability measure* on the set S of all simple events. Our hypothesis is that the observed data can be explained in terms of some probability measure.

In the case of our die-tossing experiment, we would guess that all the simple events are equally probable:

$$P(X = 1) = P(X = 2) = \ldots = P(X = 6).$$

Since

$$1 = P(S) = P(X = 1 \text{ or } X = 2 \text{ or } \ldots \text{ or } X = 6)$$

$$= P(X = 1) + \ldots + P(X = 6),$$

then the common value must be

$$P(X = 1) = \ldots = P(X = 6) = 1/6.$$

We might take this as the definition of an honest die. If we toss the die many times, we would expect the relative frequencies to be close to $1/6 = 0.16\dot{6}$.

Of course, in any experiment of this sort, we cannot expect the relative frequencies to give exactly this result. If, in 30 tosses, we were to find $f(A)$ = 6, that is, $F(A) = 0.2$, we would not be very surprised. What about if $f(A)$ = 60 in 300 tosses, or $f(A) = 600$ in 3000 tosses? How big a difference in how many tosses would be evidence that the die is dishonest? To answer this, we need to describe the independence of events in our model.

Exercises

36. If P is a probability measure on the above set S, and the probabilities of all simple events are equal, what are the probabilities of these events:

    (a) $X < 3$;

    (b) X is even;

    (c) $X < 3$ or X is even;

    (d) $X < 3$ and X is even;

    (e) X is not less than 3?

    What is the relation between $P(A)$, for any event A, and the number $N(A)$ of members in the set associated with A?

37. Consider the experiment of tossing a pair of honest dice. Let T be the total of the numbers that come up. Suppose we take the simple events to be

    $$T = k \ (k = 2, 3, \ldots).$$

    What are the possible values of T? What would be a plausible value to guess for $P(T = k)$ for $k = 2$? For $k = 7$? For any value of k? What does this give you for $P(T$ is even)? (Hint: use ordered pairs to denote simple events.)

Independence

Consider the experiment of tossing our die twice. Let X_1 be the outcome of the first toss and X_2 be the outcome of the second. The tosses are independent; that is, the outcome of either has no influence on the other. So the events

$$A: X_1 = 2$$

and

$$B: X_2 \text{ is odd}$$

should be independent. How might we test this experimentally?

We could do N times the experiment of tossing the die twice. We would obtain the relative frequencies $F(A)$, $F(B)$ and $F(A$ and $B)$. To test the influence of the event A on the event B, let us look at the $f(A)$ cases where A occurred. Among these, the event B occurred $f(A$ and $B)$ times. Thus the relative frequency of B on the *assumption* that A *occurs* is

$$f(A \text{ and } B)/f(A).$$

Since

$$F(A \text{ and } B) = f(A \text{ and } B)/N$$

and

$$F(A) = f(A)/N,$$

we have

$$f(A \text{ and } B)/f(A) = F(A \text{ and } B)/F(A).$$

Let us denote the relative frequency of B on the assumption that A occurs by $F(B/A)$. Then

$$F(B/A) = F(A \text{ and } B)/F(A).$$

If A has no influence on B, then we would expect

$$F(B/A) = F(B),$$

that is,

$$F(A \text{ and } B) = F(A)F(B),$$

at least approximately. The larger N is, the better we would expect these two quantities to agree.

This suggests that we *define* the probability of B on the assumption that A occurs (the so-called *conditional probability*) by the equation

$$P(B/A) = P(A \text{ and } B)/P(A),$$

and define A and B as independent if

$$P(B/A) = P(B),$$

that is,

$$P(A \text{ and } B) = P(A)P(B).$$

The symmetry of this last equation shows that B and A are independent if A and B are independent.

Exercises

38. Let A' be the event of 'A not occurring'. What is

$$P(A) + P(A')?$$

What is

$$P(A \text{ and } B) + P(A' \text{ and } B),$$

where B is any event?

39. Give a formula for $P(A')$ in terms of $P(A)$.

40. If A and B are independent, what is $P(A'$ and $B)$?
41. The events A, B, and C are said to be *mutually independent* if they are pairwise independent and

$$P(A \text{ and } B \text{ and } C) = P(A)P(B)P(C).$$

   Are A', B', C' also mutually independent?

The outcomes X_1 and X_2 are called *random variables* because their values depend on chance. For each k, there is assigned a probability $P(X_2) = k$ to the event $X_1 = k$. For any numbers j and k the events $X_1 = j$ and $X_2 = k$ are independent, so that

$$P(X_1 = j \text{ and } X_2 = k) = P(X_1 = j)P(X_2 = k).$$

We say therefore that X_1 and X_2 are *independent* random variables.

We can now obtain information about interesting related events. For example:

$$X_1 + X_2 = 4 \Leftrightarrow (X_2 = 1 \text{ and } X_2 = 3)$$
$$\text{or } (X_1 = 2 \text{ and } X_2 = 2) \text{ or } (X_3 = 3 \text{ and } X_2 = 1)$$

($\Leftrightarrow$ means 'is equivalent to'.) The three events occurring on the right-hand side of this equality are mutually exclusive. Hence

$$P(X_1 + X_2 = 4) = P(X_1 = 1 \text{ and } X_2 = 3) + P(X_1 = 2 \text{ and } X_2 = 2)$$
$$+ P(X_1 = 3 \text{ and } X_2 = 1)$$
$$= P(X_1 = 1)P(X_2 = 3) + P(X_1 = 2)P(X_2 = 2)$$
$$+ P(X_1 = 3)P(X_2 = 1).$$

If we assume

$$P(X_1 = k) = P(X_2 = k) = 1/6 \ (k = 1,2,3,4,5,6),$$

then we obtain

$$P(X_1 + X_2 = 4) = 3/36.$$

Exercises

42. Work out the probabilities $P(X_1 + X_2 = k)$ for all possible values of k, still on the assumption of an honest die. Compare with exercise 37.
43. Let $Y_j = 1$ if $X_j = 1$, and $Y_j = 0$ if $X_j \neq 1$, for $j = 1,2$. What does $Y_1 + Y_2$ count? Assuming the die is honest, work out the values of $P(Y_1 + Y_2 = k)$ for all possible values of k.
44. Suppose you toss an honest coin five times. What is the probability of the sequence HHTHT (H for heads, T for tails)? Let Z be the number of heads in the sequence of five tosses. Work out $P(Z = 3)$.
45. Suppose the coin is biased, so that the probability of heads in any one toss is 0.4 Now what is $P(Z = 3)$? Work out $P(Z = k)$ for $0 \leqslant k \leqslant 5$.

Expectation of a random variable

Suppose X is a random variable which can take on only the values
1,2,3,4,5, or 6. Suppose you make N independent trials, and obtain the values
$X_1, X_2, \ldots, X_N$ on these trials. Then the *average* value of X in this experiment
is

$$\langle X \rangle = \frac{X_1 + X_2 + \ldots + X_N}{N} .$$

Now the number of terms equal to 1 in the numerator is the frequency
$f(X = 1)$ of the event $X = 1$, and similarly for the other possible values of X.
Therefore we obtain

$$\langle X \rangle = \frac{f(X = 1) \times 1 + f(X = 2) \times 2 + \ldots + f(X = 6) \times 6}{N}$$

$$= F(X = 1) \times 1 + F(X = 2) \times 2 + \ldots + F(X = 6) \times 6.$$

If N is large, we expect these relative frequencies to be close to the corresponding probabilities. Hence we expect the average value of X to be close to

$$E(X) = P(X = 1) \times 1 + P(X = 2) \times 2 + \ldots + P(X = 6) \times 6.$$

This number $E(X)$ is called the *expectation* of X. In general, if X is a random
variable which can only take on the values $v_1, \ldots, v_n$ and if $P(X = v_k) = P_k$
for $1 \leqslant k \leqslant n$, then we define $E(X)$ as

$$E(X) = p_1 v_1 + \ldots + p_n v_n.$$

Notice that this is a weighted average of the values $v_1, \ldots, v_n$ with the
weights $p_1, \ldots, p_n$ (whose sum is 1).

Exercises

46. Suppose that X and Y are independent random variables, and that
    X can only take on the values 2 or 5, with respective probabilities
    p_2 and p_5 ($p_2 + p_5 = ?$), and Y can only take on the values 3, 7, or
    11, with respective probabilities q_3, q_5 and q_{11}. Give formulas for
    $E(X), E(Y)$, and $E(XY)$. Compare the latter with $E(X)E(Y)$.

47. Suppose that X and Y are as in the preceding exercise, except that
    X and Y are not necessarily independent. Let $p_{j,k} = P(X = j$ and
    $Y = k)$. Express p_2 and q_5 in terms of these *joint probabilities*.
    Compare $E(X + Y)$ with $E(X) + E(Y)$.

48. Give a formula for $E(X^2)$ in the last exercise. Compare it with $E(X)^2$.
    Try various values for p_2 and p_5. Guess the general rule. Prove it.

49. Give a formula for $E(10X)$. Compare with $10E(X)$.

Exercises 46 and 47 illustrate two important general properties of expectations:

(a) If X and Y are any random variables, then

$$E(X + Y) = E(X) + E(Y).$$

(b) If X and Y are independent random variables, then

$$E(XY) = E(X) E(Y).$$

These can be easily proved. The following properties are trivial but also basic:

(c) If X is a random variable and $X \geqslant 0$ (i.e., X can only have non-negative values), then $E(X) \geqslant 0$.

(d) If X is a random variable and c is a constant, then $E(cX) = cE(X)$.

(e) $E(1) = 1$. (Here the constant 1 on the left is thought of as the random variable which *always* equals 1.)

Exercises

50. Let A be any event, and let Y be the variable defined by the conditions

$X = 1$ if A occurs,

$X = 0$ if A does not occur.

What is $E(Y)$?

51. Find a formula for $E[(X - t)^2]$, of the form $E[(X - t)^2] = a + bt + ct^2$, where t is a constant. What is the minimum of this, as a function of t? For which value of t is it attained? Can the minimum be negative?

52. Let X be a non-negative random variable. Define Y as

$Y = 0$ if $X < 100$,

$Y = 1$ if $X \geqslant 100$.

Can $X - 100Y$ ever be negative? What is $E(X - 100Y)$? Can this number be negative?

53. Let $Y = (X - t)^2$, where t is given the value for which the minimum in exercise 51 is attained. Find an estimate for

$$P(|X - t| \geqslant 10)$$

in terms of $E(Y)$.

The law of large numbers

The reasoning of exercise 51 above leads to some important results. If X is a random variable, then $g(t) = E[(x - t)^2]$ attains its minimum only for $t = E(X)$. This minimum is non-negative, and is equal to

$$E\{[X - E(X)]^2\} = E(X^2) - E(X)^2.$$

This quantity measures the deviation of the variable X from the constant $E(X)$. It is called the *variance* of X, and its square root $\sigma(X)$ is called the *standard deviation* of X:

$$\text{var}(X) = E\{[X - E(X)]^2\},$$
$$\sigma(X) = + [\text{var}(X)]^{\frac{1}{2}}$$

(see p. 112). These are the most useful measures of the scattering of the values of X.

Suppose X is a random variable and c is any positive constant. Let Y be the variable defined by

$$Y = 0 \text{ when } |X - E(X)| < c,$$
$$Y = 1 \text{ when } |X - E(X)| \geq c.$$

Then the variable

$$Z = [X - E(X)]^2 - c^2 Y$$

is never negative. Its expectation is

$$E(Z) = \text{var}(X) - c^2 E(Y),$$

and

$$E(Y) = P[|X - E(X)| \geq c]$$

(see exercise 50). Hence we obtain

$$\text{var}(X) - c^2 P[|X - E(X)| \geq c] \geq 0,$$

or

$$P[|X - E(X)| \geq c] \leq \text{var}(X)/c^2.$$

This important result is called *Chebyshev's inequality*. If $c = 10\sigma(X)$, then

$$P[|X - E(X)| \geq 10\sigma(X)] \leq 1/100.$$

We see that it is very improbable that X differs from $E(X)$ by an amount which is large in comparison with $\sigma(X)$.

Let us look again at our die-tossing experiment. Suppose we make 100 tosses. Let X_k be the number which comes up on the kth toss. Then X_k has the same probability distribution as the variable X, considered before, which gives the number that comes up in one toss:

$$P(X_k = 1) = P(X = 1), P(X_k = 2) = P(X = 2), \text{ etc.}$$

What can we say about the average

$$Y = (X_1 + \ldots + X_{100})/100 = T/100,$$

where $T = X_1 + X_2 + \ldots + X_{100}$? We can apply the properties found above:

$E(Y) = E(T)/100,$
$E(T) = E(X_1) + \ldots + E(X_{100}).$

But since $X_1, \ldots, X_{100}$ all have the same probability distribution as X, their expectations are all the same, and

$E(T) = 100E(X),$
$E(Y) = E(X).$

This result is some evidence that our probability model is plausible. The expected value of the average result of many trials agrees with the expectation of X calculated from its probabilities.

We could now apply Chebyshev's inequality to the variable Y, but first we need to know its variance. We note that

$\text{var}(Y) = \text{var}(T)/100^2$

(see exercise 54 below). So we need to find a formula for the variance of a sum of independent random variables.

Let us try a sum of two terms first. Let

$Z = X_1 + X_2.$

Then we have

$\text{var}(Z) = E(Z^2) - E(Z)^2,$
$E(Z) = E(X_1) + E(X_2).$

We see that

$$E(Z^2) = E(X_1^2 + 2X_1X_2 + X_2^2)$$
$$= E(X_1^2) + E(2X_1X_2) + E(X_2^2)$$
$$= E(X_1^2) + 2E(X_1X_2) + E(X_2^2).$$

If X_1 and X_2 are independent, then

$E(X_1X_2) = E(X_1)\,E(X_2),$

and obtain

$$\text{var}(Z) = E(X_1^2) + 2E(X_1)E(X_2) + E(X_2^2) - [E(X_1) + E(X_2)]^2$$
$$= E(X_1^2) - E(X_1)^2 + E(X_2^2) - E(X_2)^2$$
$$= \text{var}(X_1) + \text{var}(X_2).$$

To handle a sum of three independent variables, we need only apply this result twice:

$X_1 + X_2 + X_3 = (X_1 + X_2) + X_3,$

and so on for more terms. We arrive at the general result:

The variance of a sum of *independent* random variables is equal to the sum of their variances,

or
$$\text{var}(X_1 + \ldots + X_n) = \text{var}(X_1) + \ldots + \text{var}(X_n).$$

We thus obtain
$$\text{var}(T) = \text{var}(X_1) + \ldots + \text{var}(X_{100})$$
$$= 100\text{var}(X),$$

since $X_1, \ldots, X_{100}$ all have the same distribution as X. It follows that
$$\text{var}(Y) = \text{var}(X)/100.$$

Now Chebyshev's inequality yields
$$P[|Y - E(X)| \geqslant c] \leqslant \frac{\text{var}(X)}{100c^2}.$$

In the same way, we find that if Y_N is the average of the outcomes of N independent trials then
$$P[|Y_N - E(X)| \geqslant c] \leqslant \frac{\text{var}(X)}{Nc^2}.$$

This gives us the famous law of large numbers:

If c is a fixed positive number then for very large N
$$P[|Y_N - E(X)| \geqslant c]$$
is very small.

For example, if $c = 0.001$ and $N > 10^9$,
$$P[|Y_N - E(X)| \geqslant 0.001] < \text{var}(X)/1000.$$

Thus it is very unlikely that Y_N, the average outcome of N independent trials, will differ from $E(X)$ by as much as 0.001 if N is very large. This shows that our theory explains the tendency observed in such long sequences of trials. It is strong evidence that our probability model fits reality.

The above reasoning applies to the average of any N independent, identically distributed, random variables. The estimate which we obtained for $P[|Y_N - E(X)| \geqslant c]$ is not very good; we can prove that it is much smaller, but better estimates require much more advanced mathematics.

Exercises

54. Show that, if X is a random variable and c is a constant, $\text{var}(cX) = c^2\text{var}(X)$, and $\sigma(cX) = |c|\sigma(X)$.

55. If X is the outcome of tossing one honest die, compute $E(X)$, $\text{var}(X)$, and $\sigma(X)$. Find an N such that $P[|Y_N - E(X)| \geqslant 0.01] < 0.05$. Would you trust a man whose die, in this number of trials, gave a value of 3.49 for the average outcome?

56. Define the variable Z_k by

$Z_k = 1$ if $X_k = 1$,

$Z_k = 0$ if $X_k = 0$.

What is the meaning of

$T_{100} = Z_1 + \ldots + Z_{100}$, $W_{100} = T_{100}/100$?

Compute $E(W_{100})$ and $\text{var}(W_{100})$ on the assumption of a fair die. Estimate the probability

$P[|W_{100} - E(Z_1)| \geqslant 0.03]$.

Generalize to N trials. For how big an N would you be suspicious of the man if W_N were as much as 0.2?

Probability and integration

One of the discoveries of the last fifty years is that probability and integration are essentially two languages for describing the same subject. We shall illustrate the connection between these two theories by a simple example.

Imagine an ideal 'roulette wheel' (fig. 4.7) with a spinning arrow that selects a single point on the circle. Assume that the circumference of the

Fig. 4.7

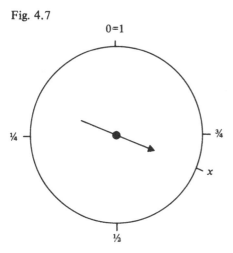

circle is 1 unit. We imagine that we can spin the arrow freely, and that when it comes to rest it will aim at some point. The selected point can be labeled by the arc-length measured from a fixed point, which would be labeled, equivalently, 0 or 1.

Let X be the outcome of a spin. Then X is a random variable which can take on any value between 0 and 1. If the spinner is perfectly honest then, by the symmetry of the circle, the probabilities

$$P(0 < X < \tfrac{1}{4}) \text{ and } P(\tfrac{1}{2} < X < \tfrac{3}{4})$$

must be equal. More generally, if $I = (a, b)$ and $J = (c, d)$ are intervals contained in $[0, 1]$ (the set of x such that $0 \leqslant x \leqslant 1$), and $d - c = b - a$, then

$$P(a < X < b) = P(c < X < d).$$

The present situation is different from those discussed previously. Before we discussed random variables which could take on only a finite number of values, but now X has an infinite number of possible values. We shall approach this case by assuming that we have concepts of probability and expectation having the properties listed in the preceding parts of this section (when we had only a finite number of possible values). We shall also use the equation

$$E(Y) = P(A),$$

where Y is defined by

$Y = 1$ if the event A occurs,

$Y = 0$ if A does not occur.

(Recall exercise 50.) Thus if

$Y = 1$ when $a < X < b$,

$Y = 0$ otherwise,

then

$$E(Y) = P(a < X < b).$$

We shall denote this random variable Y by $\chi_{(a, b)}$. Let us see where this mathematical model of the ideal roulette wheel leads us. Sometimes it will be convenient for us to cut the circle at 0 and stretch it out flat,

0   1

so that it is then represented by a line segment.

Our whole spaces of simple events now correspond to the half-open interval $[0, 1)$, the set of x such that $0 \leqslant x < 1$. Thus we have

$$P(S) = E(\chi_{[0,1)}) = E(1) = 1.$$

What is

$$P(0 \leqslant X < \tfrac{1}{2}) = E(\chi_{[0,\frac{1}{2})})?$$

Since

$$P(\tfrac{1}{2} \leqslant X < 1) = P(0 \leqslant X < \tfrac{1}{2})$$

we have

$$E(\chi_{[\frac{1}{2},1)}) = E(\chi_{[0,\frac{1}{2})}),$$

but we notice that

$$\chi_{[0,\frac{1}{2})} + \chi_{[\frac{1}{2},1)} = \chi_{[0,1)} = 1,$$

so that
$$E(\chi_{[0,\frac{1}{2})}) + E(\chi_{[\frac{1}{2},1)}) = 1.$$
This yields
$$2E(\chi_{[0,\frac{1}{2})}) = 1,$$
or
$$E(\chi_{[0,\frac{1}{2})}) = \tfrac{1}{2}.$$

Exercises

57. Evaluate
$$E(\chi_{[0,\frac{1}{3})}), E(\chi_{[0,\frac{1}{4})}), E(\chi_{[0,1/n)})$$
for any positive integer n.

58. Evaluate
$$E(\chi_{[\frac{1}{3},\frac{2}{3})}), E(\chi_{[\frac{1}{6},\frac{1}{2})}), E(\chi_{[a,b)})$$
where a and b are any rational numbers such that $0 \leqslant a < b < 1$.

59. If $b = \pi - 3$, then $b = 0.141\ 596\ 5 \ldots \in S$. Can $\chi_{[0,b)} - \chi_{[0,0.14)}$ be negative? What about $\chi_{[0,0.15)} - \chi_{[0,b)}$? Use these results to calculate $E(\chi_{[0,b)})$ to six decimal places. Calculate $E(\chi_{[0,b]})$, where $[a,b]$ denotes the set of x such that $a \leqslant x \leqslant b$, to six decimal places.

We see that in general
$$E(\chi_I) = \text{length of } I,$$
where I is any interval contained in $[0,1)$. We can use this result to estimate other expectations. For example, what is $E(X)$? A crude estimate can be obtained from the sandwich
$$0 \leqslant X \leqslant 1, \text{(hence } 1 - X \geqslant 0),$$
which gives
$$0 \leqslant E(X), 0 \leqslant E(1 - X) = E(1) - E(X),$$
so that
$$0 \leqslant E(X) \leqslant 1.$$
We can get a better estimate by noting that for $0 \leqslant X < \frac{1}{2}$,
$$0\,\chi_{[0,\frac{1}{2})} \leqslant X \leqslant \tfrac{1}{2}\chi_{[0,\frac{1}{2})},$$
and that for $\frac{1}{2} \leqslant X < 1$,
$$\tfrac{1}{2}\chi_{[\frac{1}{2},1)} \leqslant X \leqslant 1\chi_{[\frac{1}{2},1)}.$$
We can express this by the sandwich
$$0\,\chi_{[0,\frac{1}{2})} + \tfrac{1}{2}\chi_{[\frac{1}{2},1)} \leqslant X \leqslant \tfrac{1}{2}\,\chi_{[0,\frac{1}{2})} + 1\chi_{[\frac{1}{2},1)}.$$

Reasoning as before, we obtain the improved estimate

$$0 \times \tfrac{1}{2} + \tfrac{1}{2} \times \tfrac{1}{2} \leqslant E(X) \leqslant \tfrac{1}{2} \times \tfrac{1}{2} + 1 \times \tfrac{1}{2},$$

or

$$\tfrac{1}{4} \leqslant E(X) \leqslant \tfrac{1}{4} + \tfrac{1}{2} = \tfrac{3}{4}.$$

The last estimate was obtained by dividing the interval $[0,1)$ into the equal subintervals $[0,\tfrac{1}{2})$ and $[\tfrac{1}{2},1)$. Try three, four and five equal subintervals, and compare the results.

What we have done here is to estimate X by a random variable of the form

$$S = a_1 \chi_{I_1} + a_2 \chi_{I_2} + \ldots + a_n \chi_{I_n},$$

where the coefficients $a_1, \ldots, a_n$ are constants and $I_1, \ldots, I_n$ are non-overlapping intervals. When X is in the interval I_k, all terms in the above sum except the kth are equal to zero, so that

$$S = a_k \text{ when } X \text{ is in } I_k.$$

The graph of the sum S as a function of X looks like fig. 4.8. Functions of this

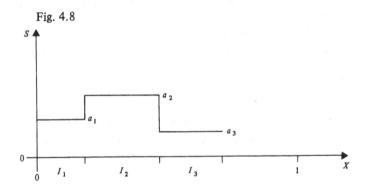

Fig. 4.8

type, which are constants in the various intervals $I_1, \ldots, I_n$ into which $[0,1)$ has been divided, are called step functions.

Thus we have found various pairs s and S, of step functions, such that

$$s \leqslant X \leqslant S.$$

From each such 'sandwich' we obtain a sandwich for $E(X)$:

$$E(s) \leqslant E(X) \leqslant E(S).$$

We can calculate the expectation of a step function explicitly, for example,

$$E(S) = (a_1 \cdot \text{length of } I_1) + \ldots + (a_n \cdot \text{length of } I_n).$$

In the same way we can compute

$$E[f(X)]$$

for any reasonable function f of X, for example, $f(X) = X^2$ or e^X.

If you examine the numbers $E(s)$ and $E(S)$, where s and S are step functions such that

$$s \leqslant f(X) \leqslant S,$$

you find that they are precisely the Riemann sums for computing the integral of f. We thus arrive at the conclusion that

$$E[f(X)] = \int_0^1 f(x)\mathrm{d}x.$$

This is the simplest case of the general connection between probability and integration.

5

Optimal solutions

Mathematics is being applied more and more frequently in the social and in the political sciences, mainly as a tool for making decisions. This often implies looking for optimal states and situations. In this chapter we explore some of the techniques used in optimalization.

The various units in this chapter are of different kinds. In section 5.1 we discuss zero-sum games, payoff matrices, and pure strategies at a level suited to grades 6-8. Mathematical models, in this case, help us to arrive at decisions.

In other sections mathematical analysis helps us to solve the problem completely: in section 5.2 written as a text for students, minimalization suggests a practical solution to a power-grid problem through graphs, spanning trees, and different minimizing algorithms; in section 5.3, the finding of maxima or minima without calculus, both at an elementary and an intermediate level, leads us also to a complete solution of the given problem.

Section 5.4 discusses the use of the least action principle for reflection and refraction in optics. The section on 'least squares' discusses 'goodness of fit', regression lines, and a geometric interpretation of the correlation coefficient. It ties in well with chapter 4 on statistics. We end this chapter with some remarks on mathematical models and decisions.

5.1   Games

We describe here an activity which we have found effective with children in grades 6-8. We start with a game between two players, Tom and Jerry. Each has a black checker and a red checker. At a signal each puts forward a closed hand containing a checker. Then they open their hands to show their choices. The payoffs are given by this array, called the *payoff* matrix:

		Jerry	
		B	R
Tom B		0	1
Tom R		−1	2

We play so that whatever one player wins, the other loses. Therefore it is enough to mention what the payoffs are for one of the players; we opt to write down Tom's payoffs. The above matrix means that if Tom chooses black and Jerry chooses red, then Tom wins one point and Jerry loses one. A payoff of −1 means a *loss* of one point. Thus if Tom chooses red and Jerry chooses black, then Tom loses one point while Jerry wins one.

After explaining the game, we play against the class. The class plays the role of Tom and we play the role of Jerry. We step out of the classroom while the class decides on their choice. An umpire is then told both our choices and writes them on the blackboard. After about five plays, we switch roles and play five more times. We usually beat the class, often with both roles. It becomes clear that we have a *strategy.*

It is then a good time to analyze the game. What is the best choice for Tom? In our experience the first thought of the children is to be optimistic. If Tom chooses red, he has a chance of winning two points, but if he always chooses red, Jerry will choose black and Tom will lose one point.

We suggest using a pessimist's approach. 'If Tom chooses black, what is his worst outcome?' Clearly it is 0. 'What about if he chooses red?' Obviously −1. 'What is the best of these worst outcomes?' Of course, 0. Therefore a pessimistic Tom will always choose black.

In the same way, we find the following results for Jerry: if Jerry chooses black his worst outcome will be 0; if he chooses red his worst outcome will be to lose 2 points. Hence Jerry's best choice is black. So with the best strategy on both sides the game is a draw. If either player always chooses black, then the other player cannot do better than also to choose black.

We can conveniently summarize the above analysis as follows:

		Jerry			
		B	R	row min	max min
Tom	B	0	1	0	0
	R	−1	2	−1	
column max		0	2		
min max		0			

Here on the right we have recorded the minimum in each row, which gives the worst outcome for Tom with each choice. Tom wants this minimum to be as large as possible, so he is interested in the *maximum of the row minima.* On the other hand, the worst that can happen to Jerry for each choice is the maximum payoff to Tom in each column. Jerry wants to find the *minimum of the column maxima.*

Both Tom and Jerry based their computations on the *worst possible out-*

comes. It is of course easier to remember this as a working rule, instead of trying to memorize who wants the minimum of the maxima of the rows/columns.

After this discussion we write the following payoff matrix on the black-board

		Jerry	
		B	R
Tom	B	1	−1
	R	0	2

and we play against the class as before. We usually win again. It is then time to analyze the new game.

If we make our table as before, we obtain

		Jerry		row min	max min
		B	R		
Tom	B	1	−1	−1	
	R	0	2	0	0
column max		1	2		
min max		1			

At first sight it seems as though Tom's best choice is red and Jerry's best choice is black. But if Jerry always chooses black then Tom can switch to black and win one point each time. However, if Tom always chooses black, Jerry can 'punish' him by changing to red. And if Jerry keeps this up, Tom can switch back to red. After a while Jerry will return to black, and so on.

Thus there is an essential difference between these two types of games. In the first case, there is a *best pure strategy*, that is, there is a certain best choice for each player to make all the time. In the second, the best strategies are *mixed strategies*: the players should mix their choices in certain ways. At this point, we ask the messenger 'What was I doing when you called me back to the room?' And he reports to the class, 'You were tossing a coin!'

The best mixed strategies consist in choosing the alternatives *at random* with certain *probabilities*. In the above game, the best strategy for Tom is to choose black and red each with probability $\frac{1}{2}$. The best strategy for Jerry is to choose black with probability $\frac{3}{4}$ and red with probability $\frac{1}{4}$. Hence when we were Tom, we tossed a coin once and chose black or red according to whether the coin came up heads or tails, but when we were Jerry, we tossed the coin twice and chose red only when we tossed two tails.

For a first introduction we have not felt it worthwhile to go beyond this point for young pupils. We have passed around the class such books as

D. Blackwell and M. A. Girshick, *Theory of games and statistical decisions* (Wiley, New York, 1954) or J. D. Williams, *The compleat strategyst, being a primer on games of strategy* (McGraw-Hill, New York, 1966); the latter is a delightful, very elementary introduction to the subject. A little general discussion of the applications to business, government and military strategy is also good for motivation.

For older students who have had a little probability and algebra, one may discuss how one finds the best mixed strategy when there is no best pure strategy. We shall not enter into this topic here.

The above examples are *zero-sum* games, that is, whatever one player wins the other loses. In other games, one must specify the payoffs to both players. In the following payoff matrix we have specified Tom's payoff first, then Jerry's:

		Jerry	
		B	R
Tom	B	(5,5)	(−4,6)
	R	(6,−4)	(−3,−3)

Thus if Tom chooses red and Jerry black, Tom wins six points and Jerry loses four. Another example of a non-zero-sum game is

		Jerry	
		B	R
Tom	B	(1,2)	(−1,−1)
	R	(−1,−1)	(2,1)

Games of this sort may be used to simulate various social situations. For example, in some games cooperative behavior is rewarded while others reward treachery.

The strategies may be affected by the possibility of the players discussing their choices beforehand, whether they can be trusted to keep agreements, whether agreements are binding, whether the games are played only once or repeatedly, and by the bargaining power of the players.

The children may find it instructive and amusing to discuss the above two games and also these:

		Jerry	
		B	R
Tom	B	(0,10)	(10,0)
	R	(−1, −20)	(−4, −30)

		Jerry	
		B	R
Tom	B	(1,2)	(7,3)
	R	(4,10)	(2,1)

		Jerry	
		B	R
Tom	B	(1,2)	(3,1)
	R	(0,−20)	(2,−30)

		Jerry	
		B	R
Tom	B	(2,−10)	(5,3)
	R	(4,4)	(3,2)

Of course we could also study games with more than two outcomes for each player to choose from. Games for more than two persons are much more difficult to analyze.

Exercises

1. Write down any payoff matrix for a zero-sum game. Analyze it. Is there a best pure strategy?
2. Discuss the above games. (In parts *a, b, c*, the game is played once only).
   (*a*) Assume no discussion in advance.
   (*b*) Assume discussion in advance, equal bargaining power, and binding agreements.
   (*c*) Assume discussion in advance, equal bargaining power, agreements not necessarily binding.
   (*d*) Assume game played repeatedly.
   There is not necessarily one best strategy. There is also disagreement, even among experts, on the answers.

5.2 Minimizing distances

Suppose that we have to connect a group of new houses to the power grid, that the local bylaws specify that power lines have to run along roads, and that we are looking for the most economical way of doing it. How do we proceed?

First of all we will of course have a look at a map (fig. 5.1) indicating the placement of houses and roads. The important factor is the *distance* between the houses, not the curvature of the roads. Therefore we can replace the map

Fig. 5.1

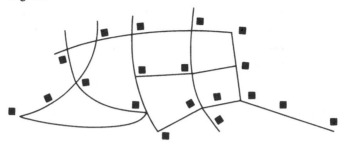

in fig. 5.1 by the systematic representation in fig. 5.2, where the number on each segment represents the distance between its endpoints. In mathematics this is called a *graph*, the points are its *nodes*, the lines its *edges*. Note that

Fig. 5.2

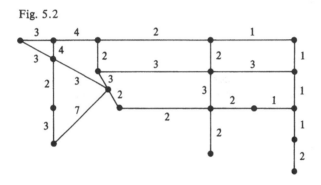

in a graph there may be isolated nodes. For instance, fig. 5.3 shows a 3-node graph.

Fig. 5.3

What we have in fig. 5.2 is a 19-node graph. It is immediately clear that if houses A, B and C are joined as in fig. 5.4 then it is not necessary to close the triangle for our purpose. Indeed, if we look for the most economical solution, we want to avoid closed paths entirely.

Fig. 5.4

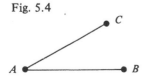

In graph theory a closed path is called a *circuit*, and a path that has no circuits a *tree*. A graph that has no isolated nodes is said to be *connected*. Which parts of fig. 5.5 are trees? A tree which joins all the nodes of a given graph is called a *spanning tree*. So, the mathematical formulation of our power-grid problem is

Find a minimal spanning tree for a given connected graph.

Fig. 5.5

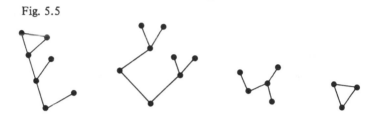

Exercises

3. Show that in a tree there is a unique path between any two nodes.
4. Try to find a minimal spanning tree for our graph (fig. 5.2), by trial and error.
5. Let us start with a 3-node graph:

Can you give a general rule for finding a minimal spanning tree in a 3-node graph?
6. How would you go about finding a minimal spanning tree for the 8-node graph in fig. 5.6? Is the solution unique? How many edges does it have?

Fig. 5.6

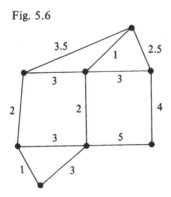

7. Show that a spanning tree of a connected graph with n nodes has $(n-1)$ edges.

A solution to exercise 6 is given in fig. 5.7. We notice that it contains all the shortest edges (with length one unit). This, together with the theorem of

Fig. 5.7

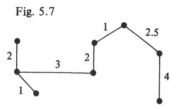

exercise 7, suggests that we could try to build up a minimal tree by *looking at the edges*, selecting the shorter ones, eliminating redundant ones, and going on until we have the necessary number of edges. For exercise 6 the first steps in the solution are shown in table 5.1.

Table 5.1

Step	State of graph	Description	Number of edges
1		Pick an edge of length one unit	1
2		Connect by taking the shortest *available* edge. (Explain *available*)	2
3		Repeat step 2.	3
4		Repeat step 2.	4
5		Repeat step 2. Be sure to examine your definition of *available edge*.	5
6		Repeat step 2.	6

At step 1 and again at step 3 we had to make a choice. Does this choice influence the solution? What we did in fact was to add, at each stage, an edge of minimal length which joined a node already considered to one not yet joined. We could formalize this procedure as follows:

(a) Start with any edge of shortest length.
(b) Call T the set of edges and nodes already joined. Add to T the shortest edge which joins a node in T to a node not yet in T. When there is a tie for the shortest edge to be added, any of the tied edges may be chosen.
(c) Continue doing this until you have $(n - 1)$ edges in T.

A set of rules for how to solve a definite problem in mathematics is called an *algorithm*. What we have just explained is called *Kruskal's algorithm*.

Exercises

8. Prove that Kruskal's algorithm yields a minimal spanning tree. One way of approaching this problem is to suppose that $T = \{e_1, e_2, \ldots, e_{n-1}\}$ is the tree we obtained, with T_1 a minimal spanning tree which has been chosen to have a maximal number of edges in common with T, and $T \neq T_1$. Let e_k be the first edge in T but not in T_1. Say the edge e_k joins the nodes A and B. Look at the path from A to B in T_1. Can it have edges shorter than e_k? Would they be included in T? If the path has an edge of length e_k, what about a substitution? What assumption would this contradict?

9. For exercise 6 the steps in a solution could have been those shown in table 5.2. What were the rules used in this solution? Formalize them! Will they always lead to a solution? (Hint: avoid circuits.) This method is called Prim's algorithm.

10. Solve our initial problem of the power grid, using Kruskal's and Prim's algorithms. Do you get the same solution both ways?

11. A traveling salesman wants to find the shortest path between A and B in the road network in fig. 5.8. It is easy to find just by inspection, but could you find an algorithm which would work in any connected graph for any pair of nodes? Start with graphs like fig. 5.9.

Table 5.2

Step	State of graph	Description	Number of edges
0			
1		Pick up all edges of length 1.	2
2		Add all edges of length 2.	4
3		Add edge of length 2.5.	5
4		Add only *one* edge of length 3. (Why?)	6
5		Add the edge of length 4.	7

Fig. 5.8

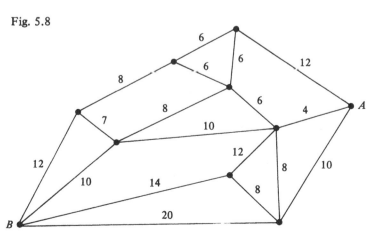

Fig. 5.9

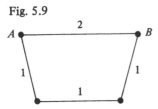

5.3 Maxima and minima without calculus

Here we consider problems where mathematical analysis permits us to list the different choices (e.g., the length of one side of a rectangular field), to predict the outcome for each choice (e.g., the cost of the fence required), and therefore to choose the best possibility.

The various optimalization problems which appear here are also useful in other ways: they provide us with good motivated problems for practice in computation and teach students how to arrange data in an orderly fashion. The standard calculus texts give many routine problems on maxima and minima. Our aim in this section is to show how some of these problems can be treated without calculus, and to show how these problems can be used advantageously in more elementary teaching.

Maxima and minima at an elementary level

The problem of maximum area for a given length of boundary can be given a twist which can be used for an interesting lesson at about the level of grades 5-6. We present the pupils with the following problem:

> Farmer Jones wants to fence off a rectangular field with an area of 10 square kilometers on his property (fig. 5.10). He invites competitive bidding for the contract. Fencing costs \$10 per km of fence.

Fig. 5.10

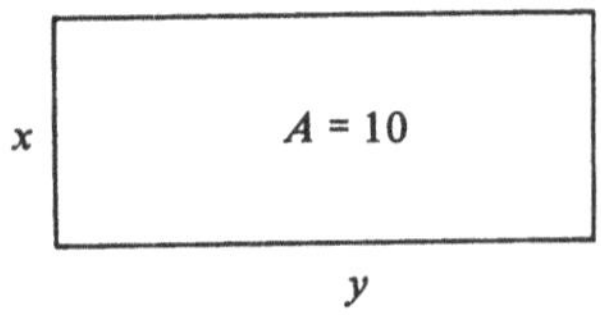

$$x \begin{array}{|c|} \hline \\ A = 10 \\ \\ \hline \end{array}$$
$$y$$

Whoever can offer to build the cheapest fence gets the job. All the bidder needs to tell is what the width and length of the rectangle should be, and how much it will cost. Thus a bidder must tell the dimensions of the fence he offers to build.

We assume that the pupils know that

area = length times width,

or

$A = xy$.

For example, if someone chooses $x = 1$, then y must be 10. Then the cost c is in dollars

$c = 10(1 + 10 + 1 + 10) = 220$.

Can anyone in the class beat that? This should stimulate some bids from the class. Usually at first the children will only try integral values for x or y. Even so, they run into calculations with fractions. Quite soon someone will try

$x = 3, y = 10/3 = 3\frac{1}{3} = 3.3\dot{3}$,

$c = 10 \times 2 \times 6.3\dot{3} = 126.67$.

After the students spend a few minutes trying to beat this value, we could, if necessary, suggest that they should try fractional values for x. We could work out a table of values for x, y, and c with them (table 5.3). This will suggest trying values of x near 3. Calculators should be available so that the

Table 5.3

x	y	c
1	10	220
2	5	140
3	$3\frac{1}{3}$	126.67
4	$2\frac{1}{2}$	130
5	2	140

pupils may try values such as $x = 3.1$ or $x = 2.9$ to see what is the trend. After some further discussion, we may say 'I think I can beat any bid you make. If you choose $x = 3$, I can beat that with $x = 3\frac{1}{6}$. Check it and see for yourselves.'

In general, if anyone bids values of x and y, we choose our width to be the average

$$\frac{x + y}{2}$$

of the pupil's width and length. We have never run into a class where someone has suggested $x = \sqrt{10}$, which is, of course, the best choice, but usually we do get to the point where the pupils notice that the fields are getting more and more square. That is, the better bids have x and y almost equal. Thus we can end on the note that if we only could make them equal, in other words if x were a number which, when multiplied by itself, gives 10 then this would be the best choice.

We could compare these results with those for the area A being four or 100 square kilometers. In these cases the students can compare the results for x = 2 or x = 10 with any nearby values and see that these seem really to be the best choices.

Maximal area and volume

With a given amount of fencing, what is the maximum area that can be enclosed? With a given amount of wood, what shape of wine barrel will contain the most wine? These are natural and practical questions. The second is related to the work of Kepler in 1615 on the solid geometry of wine barrels, which made important contributions to the development of calculus. The general problems we have stated are too hard for us to handle at this stage, but we can get some ideas by studying a few important special cases.

Suppose we only consider fences that enclose *rectangular* regions (fig. 5.11).

Fig. 5.11

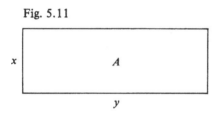

The *perimeter p* (amount of fencing) is

$$p = 2x + 2y = 2(x + y),$$

and the area A is

$$A = xy.$$

Thus we can formulate this special case of the first problem:

Given $x + y = p/2$, what is the greatest possible value of $A = xy$?

To simplify our calculations we denote $p/2$ by s:

$$s = p/2.$$

Now we can express y in terms of x:

$$y = s - x.$$

Then A is expressed in terms of just the *one* variable x:

$$A = x(s - x) = sx - x^2.$$

We see immediately that when x is close to zero or close to s then A is small. Let us graph A as a function of x (fig. 5.12). Here we have chosen $s = 6$. It looks as though x increases from 0 to 6, that A at first increases and then decreases, and that the maximum seems to be about 9, when $x = 3$. Is this really the maximum of A?

Fig. 5.12

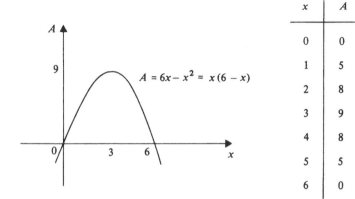

x	A
0	0
1	5
2	8
3	9
4	8
5	5
6	0

Let us check:

$$9 - A = 9 - 6x + x^2 = (3 - x)^2.$$

Now the square of any real number is non-negative, so that $9 - A \geqslant 0$, or

$$A \leqslant 9. \tag{5.1}$$

Furthermore, $(3 - x)^2$ is *positive* unless $x = 3$. Hence the equality holds in (5.1) only when $x = 3$, and then

$$y = 6 - x = 3.$$

Thus of all rectangles with a perimeter of $p = 2s = 12$ units, the one with the maximum area is the 3×3 square.

We can also give a geometric comparison. Suppose, for example, that $y > 3$. Let $d = y - 3$, so that

$$y = 3 + d, x = 6 - y = 3 - d.$$

Let us examine fig. 5.13. Here we have

$$A = (\quad) + (\quad),$$

and

$$9 = 3 \times 3 = (\quad) + (\quad).$$

Fig. 5.13

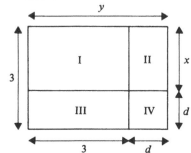

Rectangle III has the same shape as

$$(\quad) + (\quad).$$

Hence

$$9 - A = (\quad).$$

The spaces have been left for you to complete.

Exercises

12. Generalize both methods for any positive value of s.
13. Of all rectangles with area of nine square units, which one has minimum *perimeter*?

We can express the result of the preceding text in another interesting form. If $s = p/2$ is the given sum of x and y, then the square $(x = y)$ would have the side length

$$x = y = s/2,$$

so that the area of the square is $(s/2)^2$. Since this is the maximum, we obtain for any x and y

$$xy \leqslant (s/2)^2 = \left(\frac{x+y}{2}\right)^2, \tag{5.2}$$

and the equality holds only when $x = y$.

The inequality (5.2) has a very suggestive interpretation. The quantity $(x + y)/2$ is, of course, the *arithmetic mean m* of x and y. It is the number m such that the triple

$$x, m, y$$

is an *arithmetic progression*:

$$m - x = y - m.$$

The *geometric mean g* of two numbers x and y is the number such that the triple

$$x, g, y$$

is in *geometric progression*:

$$g/x = y/g.$$

(Here we consider only positive numbers.) We find that

$$g = (\quad),$$

so that (5.2) states that

$$g \leqslant m.$$

In other words, the *geometric mean of two positive numbers is never greater than their arithmetic mean*, and the two means are equal only when the numbers are equal.

Exercises

14. The arithmetic mean of n numbers is their sum divided by n. The geometric mean of n positive numbers is the nth root of their product. Show that for any *four* positive numbers $g \leqslant m$. When is $g = m$?

15. Show that the same is true for any *three* positive numbers. (Hint: use the previous exercise and choose the fourth number in a clever way.)

Our result for rectangles can be expressed in another way: of all rectangles with a given perimeter, the most symmetrical one (the square) has the largest area. This formulation suggests analogous problems for other figures, and also some good guesses as to their solutions:

> Of all triangles with a given base and perimeter, which has the largest area?
> Of all triangles with a given perimeter, which has the largest area?
> Of all quadrilaterals (of any shape) with given perimeter, which has the largest area?
> Of all closed curves with a given length, which encloses the largest area?

Another method for attacking maximum and minimum problems was discovered by Fermat in 1638. This is the basis of the approach in calculus. Let us illustrate Fermat's method with the above problem.

We wanted to find the number x which makes $A = 6x - x^2$ a maximum. Suppose that A is a maximum when $x = X$. Then for $x = X + h$, the new value of A cannot be larger. We can describe this relation by the inequality

$$6(X + h) - (X + h)^2 \leqslant 6X - X^2,$$

which is equivalent to

$$6(X + h) - (X + h)^2 - (6X - X^2) \leqslant 0.$$

If we calculate the left-hand side we obtain

$$Bh + Ch^2 \leqslant 0,$$

where

$$B = 6 - 2X.$$

What is C?

This must be true for all values of h, positive or negative. Fermat's idea is to consider values of h *very close to zero*. If h is ± 0.01 or ± 0.00001, then h^2 is much smaller than h, numerically. Therefore, *unless B is zero*, the term Bh is the larger term, and the other term will be small in comparison with it. If B is positive, the inequality is impossible for small positive values of h. If B is negative, the left-hand side will be positive for small negative values of h.

Hence B must be zero, and this yields

$X = 3.$

Thus this is the only value of x at which A *can* assume its maximum.

Exercises

16. How can you tell from the *sign* of C that $x = 3$ *does* give the maximum value of A?
17. Find the maximum of $3x - x^3$ for $x \geqslant 0$. Compare with the value when x is large and negative. Graph $y = 3x - x^3$ and explain.

Consider the problem of making a cardboard box with square base using a given amount of material (fig. 5.14). Which dimensions will give a box enclosing the greatest volume?

Fig. 5.14

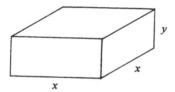

The amount of material is essentially proportional to the surface area. The boundary of the box consists of the top and bottom squares and the four rectangles on the sides. Thus the area is

$A = 2x^2 + 4xy.$

Of course, the volume is

$V = x^2 y.$

As before, we reduce the problem to the study of a function of one variable by solving the first equation for y and substituting in the second. We find that

$4V = Ax - 2x^3.$

Exercises

18. Finish this problem. What is the shape of the box of maximum volume?
19. Solve the problem for a box without a lid.

5.4 Fermat's principle in optics

It has been known for a long time that light travels in straight lines in air, water, or any other homogeneous medium, and also that it is reflected (fig. 5.15) by a mirror in such a way that the angle *i* which the incident ray makes with the perpendicular to the mirror is equal to the angle *r* which the reflected ray makes with that perpendicular. In the 1630s Snell discovered the law of

Fig. 5.15

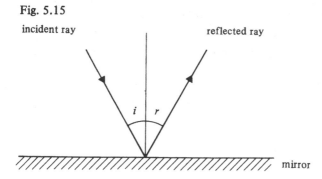

refraction, which describes how a ray of light is bent (fig. 5.16) when it passes from one medium to another, say from air to water. Snell's law states that

$$\frac{\sin i}{\sin r} = k,$$

Fig. 5.16

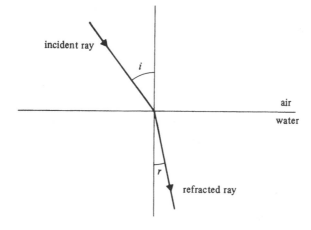

where *k* is a constant which depends only on the two media. Shortly afterwards, Fermat discovered a unifying principle by means of which he could explain these phenomena. Since then Fermat's principle has been applied to solve many other problems in optics.

Fermat (1601–1665) was by profession a lawyer, and for many years was a judge in Toulouse. He did research in mathematics, and did more in his spare time than most other people working full time on research. Fermat's principle is that light, in traveling from a point A to a point B, takes the path along which the time of travel is a minimum.

In a homogeneous medium, the velocity v of light is a constant. Hence the time it takes to travel along a path of length s is s/v. Therefore the light will travel on a path from A to B whose length s is a minimum. Thus light travels along straight lines in such a medium because a straight line is the shortest path joining two points.

Consider now light traveling from a point A to some point C on a mirror, and then to point B (fig. 5.17). We assume that the medium through which

Fig. 5.17

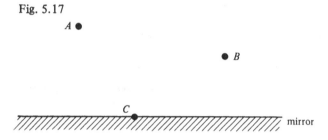

the light is traveling is homogeneous. Therefore, by our previous result, the light will travel on a straight line from A to C and then from C to B. Thus the time of travel is

$$t = \frac{AC + CB}{v} .$$

For which point C on the mirror does the broken line ACB have the shortest length?

We can solve this problem by considering the 'mirror image' B' of B with respect to the mirror (fig. 5.18). Then $CB = CB'$, so that $AC + CB = AC + CB'$.

Fig. 5.18

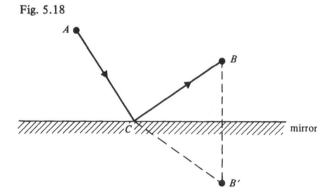

Our problem is to locate C so that the length of the broken line ACB' is a minimum. Clearly the minimum is attained when ACB' is a straight line (fig. 5.19), and we see, by elementary geometry, that the angle of incidence equals the angle of reflection.

Fig. 5.19

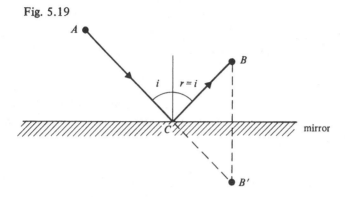

Now let us consider the problem of refraction. Assume that the velocities of light in air and water are v_1 and v_2 respectively. Choose the surface of the water as the x-axis (fig. 5.20). Suppose the light travels from A to B, and set

Fig. 5.20

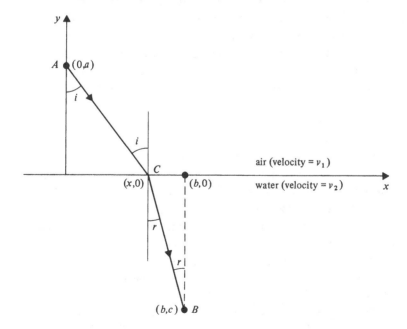

up the y-axis through A. The light will travel in a straight line from A to some point C on the surface of the water, and in a straight line from C to B. The time of travel along this path is

$$t = \frac{AC}{v_1} + \frac{CB}{v_2}.$$

Exercises

20. Express AC and CB in terms of the coordinates, as indicated in the figure.
21. Calculate dt/dx.
22. What happens at the point C for which the minimum length is attained? Interpret your result in terms of the trigonometric functions of the angles i and r.
23. How is k in Snell's law related to the velocities v_1 and v_2?

There is a similar unifying principle in mechanics called the *principle of least action*. According to this principle, a mechanical system, in passing from one state to another, takes the path for which a certain quantity, called the *action*, is a minimum.

Minimum principles like these have turned out to be powerful tools in many branches of physics and engineering. This has led the researchers in mathematical biology to search for similar principles. Some interesting principles have been proposed, but so far none has been as successful as have been the minimum principles in physics.

Long after Fermat formulated his principle, it was discovered that it is not quite correct as he stated it, but about 80 years ago it was found that the statement is still correct if the points A and B are sufficiently close to each other.

5.5 Least squares

Accounting at the X-company

The X-company manufactures plastic kitchen utensils which it sells to groceries, supermarkets, and hardware stores through a large body of salesmen. Before paying travel expenses to the salesmen, the accountants of the company check every one of their expense accounts, which entails maintaining a large staff of accountants.

The chairman of the company turned to a managerial consultant hoping that a way could be found to economize on the accounting budget. The consultant noticed that if he plotted the data of expense versus duration of trips on a graph, he obtained something like fig. 5.21. So his next step was to look for a straight line which came close to fitting the data (fig. 5.22).

Fig. 5.21

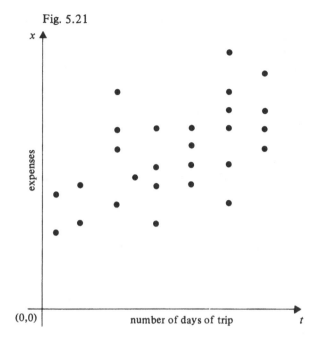

Fig. 5.22

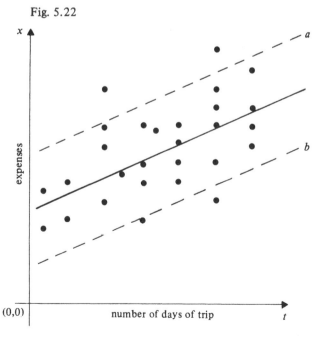

His advice was that any account which falls between the two parallels *a* and *b* to this best-fitting line should be paid without verification. Other accounts should be checked. Can you explain this advice? Do you think that it saved the company money?

The mathematical model

What the company did can be described mathematically in the following way:

(*a*) Expense accounts of the past year were plotted on a graph, the variables used being *t*, the number of days of each trip, and *x*, the cost.

(*b*) Since the plotted points seemed to cluster around a line, the 'best-fitting' line for these points was found.

(*c*) It was decided how much an account would be allowed to deviate from the 'best-fitting' line.

We will now examine (*b*) in detail. Suppose that the data we have give us the graph in fig. 5.23. We are looking for the linear relation between *x* and *t*

Fig. 5.23

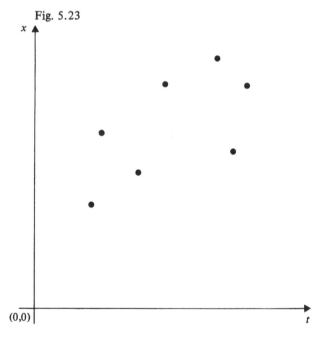

which will give us the 'best approximation' to our data. Stated this way, the problem is not well-defined. We could make our wishes more precise by stating, for example, that we want the sum of the absolute values of the distances from all points to the line to be minimal (see fig. 5.24).

Fig. 5.24

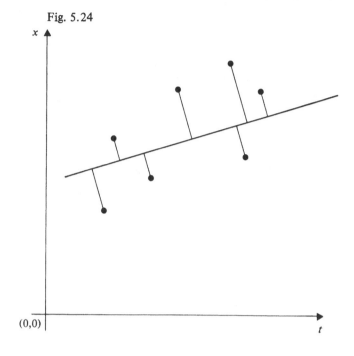

Exercises

24. Consider the problem of finding the 'best approximation' to the following data:

x	1	4	6	7
y	5	2	4	7

Choose a direction and find, by inspection, a line in your chosen direction for which the sum of the absolute values of the distances is a minimum. For how many lines in this direction is the minimum attained? Rotate your chosen direction. How does your 'best approximation' change? Look for a direction in which the approximation is best.

25. For a line of 'best approximation' in a given direction, how many of the given points are on each side of the line?

26. Add a fifth point (2,4). How are your answers to the two previous exercises changed?

27. What happens if you count distances on one side of the line as positive and distances on the other side as negative?

The regression line of x on t

In the problem of the salesmen's expense accounts, we want a straight line

$$x = at + b \qquad\qquad (5.3)$$

which gives the 'best' prediction of the value of x, given the value of t. We could interpret b as the average cost of putting a salesman on the road, and a as the average cost per day. Suppose that we are given the points (x_j, t_j), $j = 1, \ldots, n$. For the value $t = t_j$, equation (5.3) predicts the value $at_j + b$ for x. Thus the error in this prediction is

$$e_j = (x_j - at_j - b). \qquad\qquad (5.4)$$

We want to measure the 'goodness of fit' of the line (5.3) to the data by means of some function of the individual errors $e_1, e_2, \ldots, e_n$ (fig. 5.25). Of course, the function should be symmetrical with respect to these numbers.

Fig. 5.25

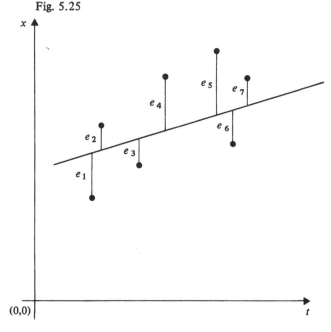

Also, negative errors should count as much as positive errors. Some of the possibilities are

$$F = \max_{1 \leqslant j \leqslant n} |e_j|,$$

$$F = \sum_{j=1}^{n} |e_j|,$$

$$F = \sum_{j=1}^{n} e_j^2,$$

$$F = \sum_{j=1}^{n} e_j^4.$$

It turns out that the algebra is the simplest for

$$F = \sum_i e_i^2 = \sum_i [x_i - at_i - b]^2. \tag{5.5}$$

As an illustration, let us find the best-fitting line thus defined for the three points $A(1,1)$, $B(3,4)$ and $C(5,4)$.

Let $x = at + b$ be the desired line; then the sum of the squares of the differences $[x - (at + b)]$ for the given points is:

$$[1 - (a + b)]^2 + [4 - (3a + b)]^2 + [4 - (5a + b)]^2$$

or

$$F = 35a^2 + 3b^2 + 18ab - 66a - 18b + 33. \tag{5.6}$$

We can minimize this expression without using calculus (as we did in section 4.4) by considering it as a quadratic polynomial.

First of all, as a polynomial in a, the minimal value of F is obtained for

$$a = -\frac{18b - 66}{70}$$

(why?), or

$$70a + 18b - 66 = 0. \tag{5.7}$$

This means in fact that, whatever the value of b, the corresponding best value for a is given by equation (5.7). Using the same reasoning on (5.6), considered as a polynomial in b, we get

$$b = (\quad). \tag{5.8}$$

Fill in the blank space here and subsequently. Therefore we have the system

$$\left. \begin{array}{l} (\quad)a + (\quad)b - 66 = 0, \\ (\quad)a + (\quad)b - 18 = 0, \end{array} \right\} \tag{5.9}$$

or

$$a = (\quad), b = (\quad). \tag{5.10}$$

The line obtained this way is called the *regression line of x on t*. When we estimate x from t using the regression equation we make the sum of the squares of the quantities e_j a minimum.

Exercises

28. Find the regression line of x on t for the following set of points:

t	1	3	5	7
x	2	5	6	9

29. Try to minimize, for the set of points given in exercise 28,
    (a) $F = \max |e_i|$.
    (b) $F = \Sigma\, e_i^4$.

The regression line of t on x

We could also have used the data (t_j, x_j) in another way, namely to estimate the length of the trip when we know the cost. To facilitate our computations we would then put t in evidence, and instead of equation (5.3) consider

$$t = \alpha x + \beta \qquad (5.11)$$

In this case we are interested in minimizing our errors when considering the quantities

$$\epsilon_j = [t_i - \alpha x_i - \beta],$$

which leads to minimizing the sum

$$F_1 = \sum_i [t_i - \alpha x_i - \beta]^2. \qquad (5.12)$$

We get a result which may be surprising at first glance – the line which minimizes (5.12) is different from the line which minimizes (5.5). This new line is the *regression line of t on x*, and on our graph (fig. 5.26) we minimize the sum of the squares of the quantities ϵ_j this time.

For the points $A(1,1)$, $B(3,4)$, $C(5,4)$ considered before, this second regression line is obtained by minimizing $\Sigma_i [t_i - \alpha x_i - \beta]^2$, or

$$[1 - \alpha - \beta]^2 + [3 - 4\alpha - \beta]^2 + [5 - 4\alpha - \beta]^2,$$

which is

$$F_1 = 33\alpha^2 + 3\beta^2 + 18\alpha\beta - 66\alpha - 18\beta + 35 \qquad (5.13)$$

The two equations in α and β are

$$\left. \begin{array}{l} 66\alpha + 18\beta - 66 = 0 \\ 18\alpha + 6\beta - 18 = 0. \end{array} \right\} \quad (5.14)$$

The solution is $\alpha = 1$, $\beta = 0$, and the regression line is given by the equation

$$t = x. \qquad (5.15)$$

Let us draw both lines on the same graph (fig. 5.27). We notice that our two lines have a common point; it is easy to see that it is the point (3,3). In fact (3,3) is the centroid of the three given points (i.e., each coordinate is the

Fig. 5.26

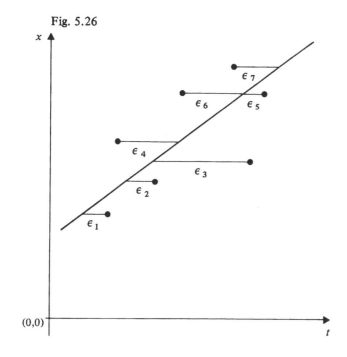

Fig. 5.27

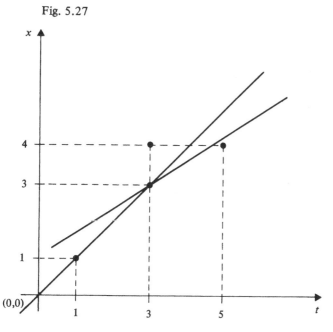

mean value of the corresponding coordinates). We will come back to this point a little later.

For students who know calculus, equation (5.9) can be obtained from the conditions that F be a minimum,

$$\frac{\partial F}{\partial a} = \frac{\partial F}{\partial b} = 0,$$

and similarly equations (5.14) can be obtained from the conditions that F_1 in (5.12) be a minimum.

More points

The same method works for more points. Let the points be (x_i, y_i), where $1 \leqslant i \leqslant n$, and suppose that we look for the line $y = mx + c$. We would then have to consider the sum of squares

$$\sum_{i=1}^{n} (y_i - mx_i - c)^2 = \sum_{i=1}^{n} y_i^2 + m^2 \sum_{i=1}^{n} x_i^2 + nc^2$$

$$+ 2mc \sum_{i=1}^{n} x_i - 2m \sum_{i=1}^{n} x_i y_i - 2c \sum_{i=1}^{n} y_i, \qquad (5.16)$$

and the system of equations to be satisfied is

$$2m \sum x_i^2 - 2 \sum x_i y_i + 2c \sum x_i = 0,$$
$$2nc - 2 \sum y_i + 2m \sum x_i = 0. \qquad \left.\right\} (5.17)$$

If we divide both sides of these equations by $2n$ and use the notation of section 4.4, we obtain

$$m \langle x^2 \rangle - \langle xy \rangle + c \langle x \rangle = 0,$$
$$m \langle x \rangle - \langle y \rangle + c = 0. \qquad \left.\right\} (5.17a)$$

The solution of this system is

$$m = \frac{\langle xy \rangle - \langle x \rangle \langle y \rangle}{\langle x^2 \rangle - \langle x \rangle^2}, \qquad (5.18)$$

$$c = \frac{\langle y \rangle \langle x^2 \rangle - \langle x \rangle \langle xy \rangle}{\langle x^2 \rangle - \langle x \rangle^2}. \qquad (5.19)$$

■   **Theorem** The y regression line on x given by $y = mx + c$, where m and c are taken from (5.18) and (5.19), passes through the centroid of the given set of points.

Prove it! Remember that the coordinates of the centroid are $\langle x \rangle$ and $\langle y \rangle$.

Exercises

30. Find the formulas similar to (5.18) and (5.19) for the regression line of x on y.
31. Prove that this regression line also passes through the centroid of the given set of points.
32. We could also have solved the above exercise in a different way, using the results of section 2.6.

   (*a*) For a fixed m, the sum in (5.16) is just like the sum in exercise 26(*b*), section 4.4. Hence, for a given m, the best choice of c in (5.16) is

   $$c = \langle y - mx \rangle = \langle y \rangle - m \langle x \rangle.$$

   (*b*) For this choice of c, the sum in (5.16) takes the form

   $$\sum [y_j - \langle y \rangle - m (x_j - \langle x \rangle)]^2,$$

   which is a function of only one variable m. Find the value of m which minimizes this sum.

   (*c*) Find the regression line of x on y by using the method of parts (*a*) and (*b*).

Measuring the correlation

   The methods we have discussed for finding regression lines apply to any set of points in the plane, but clearly in a case like fig. 5.28 no straight

Fig. 5.28

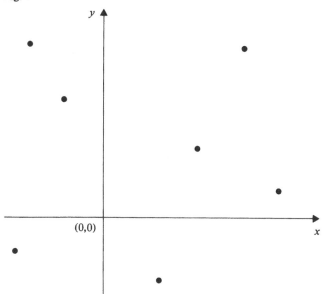

line will fit well. This leads to the problem of measuring how well a given set of data can be fitted to some straight line.

Our results suggest that the differences $x_j - \langle x \rangle$ and $y_j - \langle y \rangle$ arise naturally in our problem. This is equivalent to shifting the coordinate axes so that the new origin is at the centroid $(\langle x \rangle, \langle y \rangle)$ of the given set of points (fig. 5.29).

Fig. 5.29

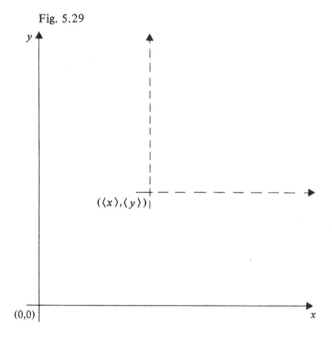

The units used in measuring x and y may not be the most natural. If we measured time in centuries, all the differences $t_j - \langle t \rangle$ in our expense account problem would be very small, and if we measured the cost x in pennies, the difference $x_j - \langle x \rangle$ would mostly be large. This suggests that we should look for some natural choice of units. The formulas (5.18) and (5.19), where the variance of x appears in the denominator, suggest that standard deviations may be such a natural choice of unit.

This leads us to introduce the *normalized variables*

$$X^* = \frac{x - \langle x \rangle}{\sigma(x)}, \quad Y^* = \frac{y - \langle y \rangle}{\sigma(y)}.$$

The value of X^* tells us by how many standard deviations x differs from its mean.

Exercises

33. (*a*) Show that

$$\langle X^* \rangle \cdot \langle Y^* \rangle = 0.$$

$$\sigma(X^*) = \sigma(Y^*) = 1.$$

(*b*) Suppose that $z = x + 5$. How is $\langle z \rangle$ related to $\langle x \rangle$? What about the relation of $\sigma(z)$ to $\sigma(x)$? How is the normalized variable

$$Z^* = \frac{z - \langle z \rangle}{\sigma(z)}$$

related to X^*?

(*c*) Let $u = 3y$. How are $\langle u \rangle$ and $\sigma(u)$ related to $\langle y \rangle$ and $\sigma(y)$? How is

$$U^* = \frac{u - \langle u \rangle}{\sigma(u)}$$

related to Y^*?

34. (*a*) Find the value of m which minimizes the sum

$$\sum (Y_j^* - mX_j^*)^2.$$

(*b*) Find the minimum of this sum.

The value of m which you found in exercise 34 is called the *correlation coefficient* (between x and y) and is usually symbolized by ρ. In terms of the original variables we can express ρ by the formula

$$\rho = \frac{\langle (x - \langle x \rangle)(y - \langle y \rangle) \rangle}{\sigma(x)\,\sigma(y)}.$$

Exercises

35. When is $\rho = +1$ or -1? (Hint: use the result of exercise 34(*b*).)
36. Give an example where $\rho = 0$.
37. Can $|\rho|$ be greater than 1?

The correlation coefficient can also be given some kind of a geometric interpretation. If we denote the slope of the regression line of y on x by m_1, the slope of the regression line of x on y by m_2, and define λ as $\lambda = m_1 m_2$, then the following exercises hint at the geometric interpretation.

Exercises

38. Prove that if n points (x_i, y_i), $i = 1, \ldots, n$, are on a straight line the value of λ is 1.
39. Can you find a set of points for which $\lambda = -1$?
40. Can you find a set of points for which $\lambda = 0$? Be careful!

41. What can you say about the two regression lines when λ is small? When λ is close to 1?
42. Prove that $\lambda = \rho^2$.

5.6   Comments on models and decisions

This is the third time we have discussed mathematical models. Previously models were considered as helping us to make *predictions*. In this chapter models have been used to help us to make *decisions*. The best way to analyze this changing role is to look carefully at more examples.

Let us look, for instance, at the question of anti-missile missiles. Both the U.S. and the U.S.S.R. are considering building such a system. The situation is that if neither of them builds such a system a stalemate exists, yet if both have it there is stalemate again, neither having gained a strategic advantage. On the contrary, both would have spent quite a large amount of money to no purpose. On the other hand, if only one of the countries acquires the system, that country has a substantial advantage, well worth the building cost.

A payoff matrix can be set up, as in section 5.1, if we put some numerical values on the different outcomes. Let us put the strategic advantage of being the the only country possessing the system at 300, the cost of developing such a system at 150; we then obtain the following matrix where the U.S.S.R.'s payoff is specified first:

		U.S.A.	
		with	without
U.S.S.R.	with	$(-150, -150)$	$(+150, -300)$
	without	$(-300, +150)$	$(0, 0)$

If you wonder what units we are using, call them national interest units.

To be on the safe side we must mention that translating different measures, such as number of deaths, crippled, or wounded, change in the measure of independence of a country, cost in money, all into these national interest units is not a scientific endeavor. These value judgements are political, and the results of our mathematical analysis will depend on them.

This matrix can now be analyzed, the virtues of collaboration (or SALT agreements) can be explained, and so can the pitfalls of insufficient enforcement of the agreements. So by forcing us to put a value on the different alternatives our mathematical model has helped us to reach a decision. This aspect of the mathematization of our problem, a careful analysis of the diverse possibilities and their relative values, is certainly a positive one but it is not the main advantage. The real importance of this model resides in the quasi-automatic reaction of the mathematician, who will look at all possible such payoff matrices.

		y
x	(a,b)	(c,d)
	(e,f)	(g,h)

study them, classify the diverse possible strategies, the cost of the various options, list the criteria that can be used, look at what happens if this is a one-time decision or one that has to be reconsidered occasionally, and by doing this make the decision makers aware of some facets of their work they might have overlooked.

Discussion problem

43. Set up a similar payoff matrix for the armament reduction problem. Can one cut down on armaments unilaterally? Would there be any incentive for the other country to do likewise?

6

Approaches to equilibrium

The main concept studied in this chapter is that of the steady state, or equilibrium, of a dynamic system. We explore several changing systems and the corresponding mathematical models. In some cases the system approaches equilibrium, in others it does not.

Investigating whether a system approaches equilibrium or not is one of the best motivations for the notions of limit and convergence. Computation of sequences and observation of their behaviour will often make a subsequent rigorous discussion of limits more natural.

Our treatment differs from the usual one in two ways. First, we deal with sequences arising in real problems, not *ad hoc* exercises. Secondly, we are not restricted to sequences defined by explicit elementary formulas.

Limits and the calculator are discussed in section 6.2. We point out how the very limitations of the calculator lead to a need for a mathematical theory. In section 6.3 we develop a more realistic model of the struggle for life. This leads to a study of quadratic equations and difference equations. In this case, as also in the sections on heat conduction (6.1) and chemical reactions (6.4), the static problem of finding the equilibrium state involves simpler mathematics than the dynamic problem of whether and how the system approaches equilibrium. By using difference equations, we can present many of these problems at an elementary level; for instance, we can discuss some aspects of heat conduction even as early as grades 5–6.

In many phenomena we encounter a resistance to disturbance of the equilibrium. With heat conduction there is a restoring velocity proportional to the deviation from equilibrium. In chapter 7 we will study cases where there is a restoring *acceleration* proportional to the deviation from equilibrium.

6.1   Heat conduction I
We give here an equilibrium problem which is suitable for pupils who

know how to compute with decimals (about grades 5-8), and which gives motivated practice with such calculations.

If you put a spoon in boiling water (fig. 6.1) the handle will gradually get warmer, but if you put it in ice-water the handle will get colder. The spoon

Fig. 6.1

boiling water   ice-water

conducts heat, and heat flows from warmer places to cooler places. Suppose you have a thin metal rod with one end in ice-water and the other end in boiling water (fig. 6.2). Imagine that you keep supplying ice at one end as it

Fig. 6.2

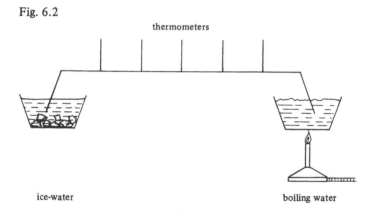

thermometers

ice-water   boiling water

melts, and boiling water at the other as it evaporates. There will be a certain temperature distribution along the rod; after a while it will be cooler near one end and warmer at the other.

At first the temperature at different places along the rod will clearly be changing with time. If you leave it in the apparatus long enough, the temperatures will no longer be changing noticeably with time. The rod will arrive at an equilibrium, or *steady state*, temperature distribution.

We wish to describe this process mathematically. Imagine the rod with thermometers placed at equal distances along it (fig. 6.3). The thermometers at the ends will always read the same: 0 °C in the ice-water, and 100 °C in the boiling water.

Consider two neighboring thermometers. If it is warmer at 4 than at 3 in

Fig. 6.3

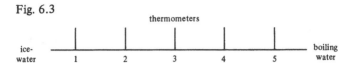

fig. 6.3, then heat will flow from the 4-point to the 3-point. Imagine that this transfers *half* the temperature difference from the warmer point to the cooler in one second. So if the readings are 50 °C and 90 °C at a particular time (fig. 6.4), heat will flow so as to transfer 20 deg. from the warmer point to the cooler in one second.

Fig. 6.4

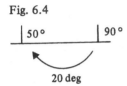

Consider a different example, of three neighboring points with temperatures 70, 50, and 90 °C at one time as shown in fig. 6.5. How will the temperature at the middle point change during the next second? Heat will flow from the two warmer points to the middle one, transferring 10 deg. + 20 deg. to it, so the

Fig. 6.5

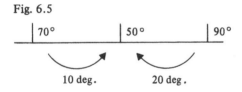

temperature will rise by 30 deg. to 80 °C. If the temperatures were 40, 50, and 90 °C respectively (fig. 6.6), then 5 deg. would move from the middle to

Fig. 6.6

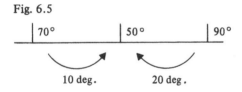

the left neighbor and 20 deg. would come from the right neighbor, so the temperature of the middle point would be

$$50 - 5 + 20 = 65 \ ^\circ C$$

after one second.

Exercises

1. Use the same values as above for the two outer points (70 and 90 °C, 40 and 90 °C), but try different temperatures at the middle point. What will be the temperature one second later at the middle point? Does it make any difference what temperature you assume at first at the middle point?
2. Try different values for the temperatures at one moment at the three neighboring points. Predict the temperature at the middle point one second later. Can you discover a simple rule for the prediction?
3. Make a table of the temperatures at the three points (table 6.1). Try various temperatures at the 2- and 4-points, and predict the temperatures one second later at the 3-point. Write in the fifth column the sum of the numbers in columns (a) and (c). Try at least five pairs of numbers of your own choice. Compare the fourth and fifth columns in table 6.1. Is there a simple relation between the numbers in these columns?

Table 6.1

Temperature (°C)			Temperature (°C) at 3-point after	
(a) 2-point	(b) 3-point	(c) 4-point	one second	(a) + (c)
70	50	90	80	70 + 90 = 160
40	50	90	65	40 + 90 = 130

We can describe the state of the rod at any moment by giving the temperatures at the 0-, 1-, 2-,..., 5-points. Thus if the whole rod is at room temperature (20 °C) initially, its state would be described by the number 20 at each point. We shall imagine that, at the ends of the rod, the temperature changes to 0 °C and 100 °C so quickly that we can describe the starting state by these numbers:

points	0	1	2	3	4	5
temperatures (°C)	0	20	20	20	20	100

In order to tabulate how the state of the rod changes with time (table 6.2), we need to make a column for the time t measured in seconds from the start of the experiment. To describe the baths of ice-water and boiling water, we put the temperatures 0 °C and 100 °C for each time in the columns for the 1- and 5-points.

Now you can begin predicting the states of the rod by using the rule you discovered in the exercises. Predict the states for the first ten seconds, calculating to the nearest whole number. Does there seem to be a trend? Calculate the states for another ten seconds. Does the trend continue? Do your predictions agree, at least *qualitatively*, with what is observed?

Table 6.2

t (time in seconds)	Temperature (°C) at					
	0-point	1-point	2-point	3-point	4-point	5-point
0	0	20	20	20	20	100
1	0					100
2	0					100
.	.					.
.	.					.
.	.					.

You may think of this as a proposed mathematical model to describe heat conduction in the rod. Since the experiment is too difficult for youngsters to do with enough precision, at this level we can at most try to make the model plausible by comparing the model qualitatively with the observed phenomenon. We cannot hope for a good quantitative agreement.

Exercises

4. Do the above calculations a little more exactly, say to one decimal place. Does the result agree with what you found before? Do you notice anything that escaped you in the less exact calculation?

5. Try other starting states and make the same prediction. Do you always seem to get an approach to a steady state? Does the steady state depend very much on the starting state?

6. Try rods of other lengths, that is, with different numbers of equally distant points. Do you get similar results? Does there seem to be any simple rule for guessing the steady state?

7. In the above model we assumed that *half* the temperature difference at neighboring points is transferred from the warmer to the cooler point in one second. Suppose only one-third of the temperature difference is transferred. Would this represent a metal which conducts heat better or worse than our original rod? Would you expect the approach to a steady state to be faster or slower? Try calculations with the same starting states you used before and compare with your previous results.

8. Guess what would happen if one-tenth of the temperature difference is transferred each second. Calculate some rod states, and check your guess.

6.2 Teaching the limit concept

The concept of limit is difficult to teach for several reasons. The most important reason is that this is the first essentially infinitary process which students encounter. Thus it marks the point of separation between elementary mathematics, dealing with finite processes of algebra and geometry, and higher mathematics. A second reason is that the study of limits requires the use of inequalities. In present curricula students often get inadequate practice in working with inequalities. In high school the presentation is usually unmotivated and accompanied by quite artificial exercises. When limits are introduced in calculus, the teacher is usually pressed for time to get to the techniques of calculus, and has no time for a digression on inequalities.

Since the most natural use of inequalities is in the study of approximations and the estimation of errors, it is clear that numerical analysis provides the best motivation for working with inequalities. That is why, in connection with each of the simpler functions, we have stressed approximation and estimation of the errors involved.

It is probably best to begin with problems where the convergence and the value of the limit are both quite obvious. Then the only problem is to find an estimate for the error and to determine when the error becomes and remains less than a prescribed tolerance. Good examples of this type of problem arise naturally in studying the relative sizes of various common functions of n.

For instance, let us compare the growth of n^k with that of a^n as n increases, where $k > 0$ and $a > 1$. Each student can choose values for k and a and compute a table for some simple functions of n. Table 6.3 gives an example for $k = 2$, $a = 1.1$. The students observe that z_n ultimately begins to decrease, and then

Table 6.3

n	$x_n = n^2$	$y_n = 1.1^n$	$z_n = x_n/y_n$
0	0	1	0
1	1	1.1	0.909
2	4	1.21	3.306
3	9	1.331	6.762
4	16	1.464	10.928
⋮	⋮	⋮	⋮
20	400	6.727	59.457
21	441	7.400	59.593
22	484	8.140	59.457
⋮	⋮	⋮	⋮
100	10 000	13 780.612	0.726

becomes very small. At which point does z_n begin to decrease? From which point on does z_n become less than 10^{-6}?

After the experience with actual numbers, it is natural to begin the theoretical investigation by asking when is z_{n+1}/z_n less than 1? We find that

$$z_{n+1}/z_n = \left(1 + \frac{1}{n}\right)^2 / 1.1.$$

Clearly this approaches $1/1.1 = 0.909$ as n increases, so that ultimately $z_{n+1}/z_n < 1$ and $z_{n+1} < z_n$, that is, z_n is decreasing. Indeed

$$z_{n+1}/z_n < 1$$

if

$$1 + \frac{1}{n} < (1.1)^{\frac{1}{2}} = 1.0488,$$

or

$$1/n < 0.0488$$

or

$$n \geqslant 1/0.0488 = 20.5,$$

which agrees with table 6.3.

Exercises

9. (a) Find a value of N such that

$$z_{n+1}/z_n < 0.91 \text{ for } n \geqslant N.$$

 (b) If n is greater than N, which is larger, z_n or $(0.91)^{n-N}z_n$?
 (c) Find a number M such that $z_n < 10^{-6}$ for $n > M$.
 (d) Find a number M such that $z_n < 10^{-10}$ for $n > M$.

10. Choose some other values for k and a.
 (a) What is

$$\lim_{n \to \infty} (z_{n+1}/z_n)?$$

 (b) Find an N such that $z_{n+1}/z_n < 1$ for $n \geqslant N$.
 (c) Find an r $(0 < r < 1)$ and an N such that $z_{n+1}/z_n < r$ for $n \geqslant N$.
 (d) Find an M such that $z_n < 10^{-6}$ for $n > M$.
 (e) Check your results in parts (a)–(d) by computation.

11. Compare the growth of a^n $(a > 1)$ with that of $n!$.
 (a) Choose a value for a and compute a table of

$$u_n = a^n/n! .$$

 (Hint: note that $u_{n+1}/u_n = a/(n + 1)$. Thus it is easy to compute u_{n+1} from u_n. How does this procedure compare with direct computation of u_n, in number of steps, overflow of calculator capacity, and roundoff?)
 (b) Find an N such that u_n is decreasing for $n > N$.
 (c) Find an M such that $u_n < 10^{-6}$ for $n > M$.

12. Compare the growth of $n!$ with that of n^n. Let $v_n = n!/n^n$.
    (a) Try to find a simple formula for v_{n+1}/v_n.
    (b) Compute a table of v_{n+1}/v_n. How does this ratio behave for large n?
    (c) According to your table, is v_{n+1}/v_n increasing or decreasing? What about v_n?
    (d) What is

    $$\lim_{n \to \infty} v_n = L \,?$$

    Find a k such that $n^k v_n \leqslant 1$ for all $n > 0$. Find an N such that $v_n < 10^{-6}$ for $n > N$.

13. Compare the growth of $\log_2 n = I(n)$ (chapter 2) with that of n^k, $k > 0$.
    (a) Choose a value of k and compute a table of $w_n = I(n)/n^k$. Guess the value of $\lim_{n \to \infty} w_n$.
    (b) For any n, let m be the integer defined by

    $$2^m \leqslant n < 2^{m+1}.$$

    Obtain an estimate of w_n in terms of m. Use exercise 10 to find an N such that $w_n < 10^{-6}$ for $n > N$.

14. Compare the growth of 2^n with 3^n. Find an N such that

    $$(2^n/3^n) < 10^{-6}.$$

    for $n > N$. Check by computation.

It is desirable to build up an intuitive feeling for the sizes of numbers and the orders of magnitude of the most common functions of n. It is more important that students know that, for any positive constant k and any constant $a > 1$, the elementary functions can be arranged in a scale

$$\log n \ll n^k \ll a^n \ll n! \ll n^n$$

(where $f(n) \ll g(n)$ means that $f(n)/g(n)$ approaches zero as $n \to \infty$), than for them to master the ϵ–δ techniques of the theory of limits. Most of the problems on limits found in the usual textbooks can be solved by inspection with the use of this scale.

In the previous exercises we have indicated how the calculator is useful in comparing the growth of these functions. The students should also look at other sequences which illustrate what sorts of things can occur. The sequences $\{x_n\}$ defined by

$$x_n = \frac{n}{n+1} \text{ or } x_n = 1/2^n$$

exhibit approach to a limit in a fixed direction. The cases

$$x_n = (-1)^n/n \text{ or } x_n = (\sin n)/n$$

illustrate oscillation around the limit. These should be contrasted with sequences like

$$\{n\}, \quad \{(-1)^n\}, \quad \text{or} \quad \{(-1)^n + (-1)^{(n^2 - n)/2}/n\},$$

which diverge in various ways.

The calculator is also useful in studying algorithms for numerical approximation. It is best to start with concrete problems such as the computation of $\sqrt{3}$ by applying the process of section 1.6 to the polynomial $P(x) = x^2 - 3$. If we start with $x_0 = 2, x_1 = 1$, then the algorithm leads to the sequence defined by

$$x_{n+1} = x_0 - \left(\frac{x_n - x_0}{P(x_n) - P(x_0)}\right) P(x_0) \tag{6.1}$$

where $n > 0$, so that

$$x_{n+1} = 2 - \frac{1}{x_n + 2} . \tag{6.2}$$

In this case the quantity $\delta_n = 3 - x_n^2$ gives a good natural measure of the error in the approximation. The students can compute the numbers x_n and δ_n for $n \leqslant 20$. It is very plausible that x_n approaches $\sqrt{3}$. A graph makes the convergence almost obvious (fig. 6.7).

Exercises

15. Compute x_n and δ_n for $n \leqslant 20$ and make a table. Does it seem that x_n is converging? Does δ_n seem to be approaching zero?

16. What difference does it make in your computations if you use (6.1) rather than (6.2)? Explain.

17. Choose any x_0 and x_1 such that $0 < x_1 < x_0, x_1^2 < 3 < x_0^2$, and compute x_n and δ_n for $n \leqslant 20$.

18. Use (6.2) to obtain a formula for δ_{n+1} in terms of δ_n. If $x_n > 0$, what is the least that x_{n+1} can be? If $x_n > 0$ and $\delta_n > 0$, can δ_{n+1} be negative? Can δ_{n+1}/δ_n be more than $\frac{1}{4}$? Find an estimate for δ_n in terms of δ_1. Find an N such that $\delta_n < 10^{-6}$ for $n > N$.

19. Let $P(x) = x^3 - x - 1$. Find x_0 and x_1 such that $0 \leqslant x_1 < x_0, P(x_1) < 0 < P(x_0)$. Apply the process of section 1.6 to find a sequence $\{x_n\}$ of approximations to the positive root of $P(x) = 0$. Compute x_n and $P(x_n)$ for $n \leqslant 20$. Does it seem as though x_n is converging to the desired root? See how the process looks on a graph.

Fig. 6.7

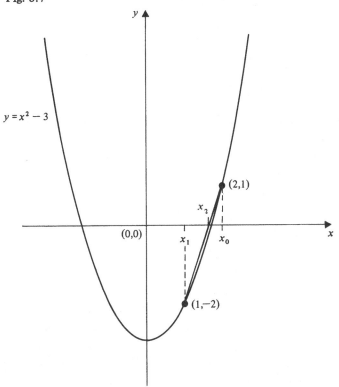

$y = x^2 - 3$

(2,1)

x_2

(0,0)

x_1 x_0

x

(1,−2)

Often we can tell without computation what the possible limits of a sequence can be. This may tell us what to look for when we compute the sequences. Thus if the sequence defined by (6.2) above converges,

$$\lim_{n \to \infty} x_n = x,$$

then also

$$\lim_{n \to \infty} x_{n+1} = x,$$

and hence

$$x = 2 - \frac{1}{x+2} \ , \text{ or } x^2 - 3 = 0.$$

Thus x must be a square root of 3, and $\delta_n = 3 - x_n^2$ is a natural measure of the error. Similarly, if

$$x_{n+1} = \frac{1}{1+x_n} \quad \text{for } n \geqslant 0, \tag{6.3}$$

then the only possible limits of x_n are the roots of the equation

$$x = \frac{1}{1+x}, \text{ or } x^2 + x - 1 = 0,$$

and $\delta_n = x_n^2 + x_n - 1$ is a good measure of the error.

In problems of this type the limit may depend on the initial values, and sometimes the dependence is not obvious.

Exercises

20. Try various values of x_1 in (6.2) above, and compute x_n for $n \leqslant 10$. Can you find values of x_1 for which x_n converges to $-\sqrt{3}$, and values for which x_n diverges? Graph the function

$$y = 2 - \frac{1}{x+2},$$

and use this to explain your numerical results.

21. Study the sequences defined by

$$x_{n+1} = \left(x_n + \frac{5}{x_n}\right) \Big/ 2, \text{ for } n \geqslant 0.$$

   (a) What are the possible limits of such a sequence?

   (b) Choose various values for x_0 and compute x_n for $n \leqslant 10$. Does x_n seem to converge? Does the limit depend on the choice of x_n? Can you find an x_0 for which the sequence diverges?

   (c) Explain your results in part (b) by means of the graph of the function

$$y = \left(x + \frac{5}{x}\right) \Big/ 2.$$

   (d) Construct a good measure for the error. Compute this measure in the cases you studied in part (b). Does this agree with your previous observations on convergence?

   (e) Find a formula for the measure of error at the $(n+1)$th step in terms of the measure at the nth step. Find an N such that the measure of error is less than 10^{-6} for $n > N$.

22. Study the sequence defined by

$$x_{n+1} = -\frac{2x_n^3}{100 - 3x_n^2}.$$

   (a) What are the possible limits of x_n?

   (b) Choose various values for x_0 and compute x_n for $n \leqslant 10$. Does x_n seem to converge? If so, how does the limit depend on the choice of x_0? Can you find a value of x_0 for which the x_n diverges? How can you explain your observations? In this case, a computer may be more useful than a calculator. In a certain sub-interval of the interval $0 \leqslant x \leqslant 10$ the dependence of behavior on x_0 is so delicate that roundoff may produce misleading results.

In the study of sequences like these where the convergence is obvious or plausible and the limit is obvious or easy to guess, the usual definition of limits is meaningful. We say that

$$\lim_{n \to \infty} x_n = x$$

if, for every positive ϵ, there is an N such that

$$|x_n - x| < \epsilon \text{ for } n > N.$$

The quantity ϵ represents the size of error, or tolerance, which we are willing to allow, and N represents how far we must go in the sequence before the error will remain below the given tolerance. Since N usually depends on ϵ, we often write $N(\epsilon)$ to emphasize this. The way $N(\epsilon)$ increases as ϵ decreases indicates how fast the sequence converges.

In order to apply this definition, we must know or guess the value of x. Since immature students usually do not appreciate the value of proofs, they do not see the point of all the fussing with estimates when they are sure of what x is. As we have suggested, when $\{x_n\}$ arises from a process for computing x, it does make good practical sense to ask how many steps are needed to make the error less than a given amount.

In cases where the value of x is not easy to guess, the above definition has no practical value. Let us look at a few cases of this type.

The sequence $x_n = (1 + 1/n)^n$ for $n \geqslant 1$ arose in section 3.1 (where $x_n = C(1, h)$ for $h = 1/n$), and also in exercise 12 (*b*) of this section. Computation of x_n for moderate values of n suggests that x_n is increasing and approaching a limit. If you compute x_n with a calculator, you may obtain mysterious results for large n. We obtained the results shown in table 6.4 (remember that $\lim (1 + 1/n)^n = 2.718\,281\,8\ldots$). Of course, the curious trend for $n = 60\,000$ and $n = 70\,000$ is due to the roundoff error.

Table 6.4

n	x_n
50 000	2.718 269
60 000	2.718 222
70 000	2.718 222
80 000	2.718 281

At the high school or junior college level, it is probably sufficient to make plausible the convergence of x_n on the basis of numerical evidence. For a more rigorous discussion at this level it is easier to deal with the subsequence defined by

$$y_n = x_k, \text{ where } k = 2^n,$$

and it is useful to compare with the sequence

$$z^n = \frac{1}{(1 - 1/2^n)^{2n}} \, .$$

Note that $y_n = C(1,1/2^n)$ and $z_n = C(1,-1/2^n)$. From the inequality

$$(1 + h)^2 > 1 + 2h \text{ for all } h,$$

we obtain, on setting $h = 1/2^{n+1}$,

$$y_n < y_{n+1}$$

and, on setting $h = -1/2^{n+1}$,

$$z_{n+1} < z_n.$$

Also we see that

$$y_n/z_n = \left(1 - \frac{1}{2^{2n}}\right)^{2^n} < 1,$$

so that

$$y_n < z_n.$$

Thus y_n is increasing and z_n is decreasing. All the ys are less than all the zs, for if $n < m$, then

$$y_n < y_m < z_m < z_n.$$

In particular, we have $y_n < z_1 = 4$ for $n \geqslant 1$. We now wish to evaluate the difference

$$z_n - y_n = z_n \left(1 - \frac{y_n}{z_n}\right),$$

and thus need to find lower limits for y_n/z_n. We use the inequality

$$(1 + h)^k \geqslant 1 + kh \text{ for } h \geqslant -1,$$

which is easy to prove by induction, and thus obtain

$$y_n/z_n \geqslant 1 - \frac{2^n}{2^{2n}} = 1 - \frac{1}{2^n} \, ,$$

which yields

$$z_n - y_n < 4/2^n = 1/2^{n-2}.$$

Thus for $n < m$,

$$0 < y_m - y_n < z_n - y_n < 1/2^{n-2}.$$

so that all the y_n, for large n, *agree with each other* to within a very small error; for example,

$$0 < y_m - y_n < 2^{-20} < 10^{-6}$$

for $m > n \geqslant 22$.

It is now plausible that y_n is converging to some limit y, and that y_{22} is an approximation to y with an error of less than 10^{-6}. From the practical, or engineering, point of view this gives a satisfactory way of calculating y to within an arbitrarily small error. Of course, y is the familiar constant e. This is the significance of the Cauchy criterion for convergence:

Given $\epsilon > 0$, there is an $N = N(\epsilon)$ such that $|y_m - y_n| < \epsilon$ for $m,n > N$.

After experience with the above sequence, the students can probably guess on the basis of computations, say for $n \leqslant 20$, that the sequence

$$u_n = \left(\frac{n^n}{n!}\right)^{1/n} = \frac{n}{(n!)^{1/n}}$$

converges, and perhaps even guess at the value of the limit of u_n.

A very instructive example is to have each student toss a coin and define the sequence $\{s_n\}$ by

$$s_0 = 0,$$

$$s_n = s_{n-1} + \frac{\sigma_n}{10^n},$$

where $\sigma_n = +1$ if the nth toss is heads, and $\sigma_n = -1$ if the nth toss is tails. Thus each student obtains a different sequence, which he can easily work out for himself, and each sequence appears to be converging. In this case, it is impossible to predict the value of $s = \lim s_n$, but we see that for $m > n$

$$|s_m - s_n| = \left| \frac{\sigma_{n+1}}{10^{n+1}} + \ldots + \frac{\sigma_m}{10^m} \right|$$

$$\leqslant \frac{1}{10^{n+1}} + \ldots + \frac{1}{10^m}$$

$$\leqslant \frac{1}{10^{n+1}} \cdot \frac{\left(1 - \frac{1}{10^{m+1-n}}\right)}{1 - \frac{1}{10}}$$

$$< (1/9) 10^{-n}.$$

Hence we can predict that each student will certainly obtain a convergent sequence.

The sequence $\{t_n\}$ defined similarly by

$$t_0 = 0,$$

$$t_n = t_{n-1} + \frac{\sigma_n}{n},$$

where σ_n is as before, is also interesting. Each student will obtain a different sequence, and all will seem to be convergent. For very large n, both the calculator and the computer will have trouble with roundoff unless special tricks are used. Although logically it is possible that t_n will diverge, this occurs only when there is such a large difference between the frequencies of heads and tails that its occurrence would be a miracle! Most of the values of $t = \lim_{n \to \infty} t_n$ lie in the interval $|t| \leqslant 1.3$, but there is a small positive probability that $|t| > 10^6$.

The Cauchy criterion for convergence should be made intuitively plausible in an elementary calculus course. At a more advanced level it must be taken essentially as a postulate, or as part of the definition of the real number system.

Another criterion for convergence, which does not require knowing or guessing the value of the limit, is:

> If x_n is non-decreasing and bounded, that is, $x_n \leqslant x_{n+1}$ for all n, and there is a B such that $x_n \leqslant B$ for all n, then $\lim x_n = n$ exists.

This can also be made intuitively plausible by marking the points on a number line. There is no constructive proof, and there is no general method for obtaining an estimate for the error $|x - x_n|$. It can be proved, by *reductio ad absurdum*, using the Cauchy criterion. A rigorous discussion should certainly be left to an advanced course.

Exercises

23. (a) Compute $x_n = 2^{1/n}$ for $1 \leqslant n \leqslant 20$. Does x_n seem to be converging? Can you guess the limit?

    (b) Use $\log x_n$ to obtain an estimate for the error. Find an N such that x_n differs from the limit by less than 10^{-10}. (Hint: use the results of chapter 3.)

24. (a) Compute $x_n = n^{1/n}$ for $1 \leqslant n \leqslant 20$. Does x_n seem to converge? Guess the limit $x = \lim_{n \to \infty} x_n$.

    (b) Find an N such that $|x_n - x| \leqslant 10^{-10}$ for $n > N$.

25. (a) Compute $x_n = (2^n + 3^n)^{1/n}$ for $1 \leqslant n \leqslant 20$. Does x_n seem to converge? Guess the limit $x = \lim_{n \to \infty} x_n$.

(b) Which is larger, 2^n or 3^n? Why does x_n approach x? Find an N such that $|x_n - x| < 10^{-10}$ for $n > N$.

26. (a) Let $f(x) = \log(1 + x)$. What is

$$\lim_{h \to 0} \frac{f(h) - f(0)}{h} \ ?$$

Apply this to find the limit of $\log[(1 + 1/n)^n]$ as $n \to \infty$.

(b) Evaluate the integral

$$J(h) = \int_0^1 \frac{dt}{1 + ht} \ .$$

Find constants A and B such that

$$Ah - Bh^2 \leqslant J(0) - J(h) \leqslant Ah.$$

for $h \geqslant 0$.

(c) Use the inequality (see chapter 3)

$$x \leqslant e^x - 1 \leqslant e^x x$$

for $x \geqslant 1$ to obtain an estimate for $J(0) - J(h)$ for $h \geqslant 0$, and to estimate

$$e - \left(1 + \frac{1}{n}\right)^n \ .$$

(d) What is

$$\lim_{n \to \infty} n \left[e - \left(1 + \frac{1}{n}\right)^n\right] \ ?$$

Check by computation. How far do you have to go before round-off is significant?

27. (a) Compute

$$x^n = y_n^{1/n} \ ,$$

where

$$y_n = \frac{1 \times 3 \times 5 \times \ldots \times (2n + 1)}{n!}$$

for $1 \leqslant n \leqslant 20$. Does it make any difference with your calculator whether you compute y_n by calculating numerator and denominator separately and then divide, or by calculating the product

$$\left(\frac{3}{1}\right) \times \left(\frac{5}{2}\right) \times \left(\frac{7}{3}\right) \times \ldots \times \left(\frac{2n + 1}{n}\right) \ ?$$

Does x_n seem to converge? Can you guess the limit?

(b) Is x_n greater or less than the guessed limit? Prove your statement.

(c) What is

$$\lim_{n \to \infty} y_{n+1}/y_n = c?$$

How does this help you understand why x_n should converge? Find an N such that

$$\left| \frac{y_{n+1}}{y_n} - c \right| < 10^{-7}$$

for $n > N$. Find an M such that $x_n - c < 10^{-6}$ for $n > M$.

28. (a) How fast does ne^{-n} approach zero as n increases? For $k = 0,1,2,$ $3,\ldots,10$, find the smallest n such that

$$n\, e^{-n} \leqslant e^{-k}.$$

Make a table of n as a function of k, and of n/k. Does n/k seem to be approaching a limit?

(b) Make a table of $n - k$ for $0 \leqslant k \leqslant 10$. Can you guess the order of magnitude of this difference? Compare it with some familiar functions of k.

(c) For $y \geqslant 1$, consider the solution of the equation

$$x - \log x = y,$$

such that $x \geqslant 1$. (Why does the solution exist? What is the sign of dy/dx for $x > 1$?) Which is larger, x or y? Which is larger, x or $y + \log y$? Set $x = y(1 + h)$, and use the inequality $\log(1 + h) < h$ (chapter 3) to get a good estimate for x in terms of y.

(d) Use the results of part (c) to improve your results in part (b).

29. Use the results of exercise 27(c) to investigate the rate of convergence of $(\log n)/n$. (Hint: set $x = \log n$, $n = e^x$.) Find approximately the smallest n such that

$$(\log n)/n < e^{-1\,000\,000}.$$

30. Consider the series

$$s_1 = \frac{1}{1} + \frac{1}{3} + \frac{1}{5} + \ldots,$$

and

$$s_2 = \frac{1}{2} + \frac{1}{4} + \frac{1}{6} + \ldots$$

(a) Toss a coin many times. Let

$$x_0 = 0,$$

and let

$$x_n = x_{n-1} + t_n,$$

where t_n is the first unused term of s_1 if the nth toss is heads, and minus the first unused term of s_2 if the nth term is tails. Thus if your first tosses are HTHTTHHTTT, then

$$x_{10} = \frac{1}{1} - \frac{1}{2} + \frac{1}{3} - \frac{1}{4} - \frac{1}{6} + \frac{1}{5} + \frac{1}{7} - \frac{1}{8} - \frac{1}{10} - \frac{1}{12}.$$

Compute and tabulate x_n for $n \leqslant 20$. Does x_n seem to be converging? Compare with the results of your classmates. Do you seem to be getting the same limits?

(b) Note that every sequence you obtain in this way is obtained by adding the terms in $s_1 - s_2$ in a certain order. Do different orderings give noticeably different results?

(c) Give methods of choosing the terms from s_1 and s_2 so that x_n approaches 100 or so that x_n oscillates between -100 and $+100$ infinitely often.

Intuitively speaking, this exercise shows that the commutative law of addition does not, in general, hold for infinite series.

31. Choose any positive numbers x_0, x_1, and x_2, and define

$$x_{n+3} = 3x_{n+1} + 4x_n \quad \text{for } n \geqslant 0,$$

and let

$$x_{n+1} = x_{n+1}/x_n \quad \text{for } n \geqslant 0.$$

(a) Compute a table of x_n and y_n for $n \leqslant 20$. Does y_n seem to be converging?

(b) If

$$\lim_{n \to \infty} y_n = y$$

exists, what must y be? (Hint: express x_{n+1} and x_{n+3} in terms of x_n and the ys.) Use this to devise a good measure of the error of $y_n - y$. Compute this measure of error and test your guess about the convergences.

(c) Does y_n seem to be converging in a fixed direction, or is it oscillating around the limit?

(d) Choose any positive number $z_0 > 1$, and define

$$z_{n+1} = \frac{2}{3}\left(\frac{z_n^3 + 2}{z_n^2 - 1}\right) \quad \text{for } n \geqslant 0.$$

Compute z_n for $n \leqslant 20$. Does z_n seem to be converging? How does its limit seem to be related to y? Which sequence seems to be converging faster?

32. Consider the sequence $\{x_n\}$ defined by (6.2). Let $\alpha = \sqrt{3}$. Use the fact that $\pm\alpha$ satisfies the equation

$$x = 2 - \frac{1}{x + 2},$$

to obtain a constant k such that

$$\frac{x_{n+1} - \alpha}{x_{n+1} + \alpha} = k \left(\frac{x_n - \alpha}{x_n + \alpha}\right).$$

(a) Find

$$\lim_{n \to \infty} \frac{x_{n+1} - \alpha}{x_n - \alpha}.$$

(b) Find

$$\lim_{n \to \infty} \frac{\log|x_n - \alpha|}{n}.$$

(c) Find

$$\lim_{n \to \infty} \frac{x_n - \alpha}{k^n}.$$

(d) How do these results explain the exact rate of convergence of x_n?

33. Study the sequence $\{x_n\}$ of exercise 21 by the method of exercise 31.

(a) Find

$$\lim_{n \to \infty} \frac{x_{n+1} - \alpha}{(x_n - \alpha)^2},$$

where α is the positive root of the equation

$$\alpha = \frac{1}{2} \left(\alpha + \frac{5}{\alpha}\right).$$

(b) Find

$$\lim_{n \to \infty} \frac{\log|x_n - \alpha|}{2^n} = \log k.$$

(c) Find

$$\lim_{n \to \infty} |x_n - \alpha|/k^{2^n}.$$

6.3   The struggle for life II

A better model

We continue here the study we started in chapter 3 of the mathematical theory of the struggle for life.

We concluded that our model was not realistic enough. As in most other problems in which we try to apply mathematics to the real world, we find that the real world is too complicated for our poor feeble human minds to grasp. Hence we try to idealize and simplify the actual situation until we obtain something easy enough for us to handle. We try to pick out the most important features of the real problem and incorporate them into a mathematical model. We often start out with a very simple mathematical model. After we have studied it thoroughly and can understand this first approximation to the real world, we then, step by step, introduce new ideas to make our model more realistic.

This is what we shall do now. Our previous model assumed a certain basic relative rate of excess of births over deaths, and that this basic rate is constant. As a first step to improve this, let us assume a correction which takes into account the rate at which the population uses up its food supply and poisons its environment. Let us assume that this correction is proportional to the size of the population.

We can express our assumption in mathematical language. Previously we assumed that r, the relative rate of change of the population, is constant. Now we are assuming that r depends on the size x of the population:

$r = R - cx$.

Here R is the basic rate of excess of births over deaths, and cx is a correction proportional to the size of the population. We assume that both R and c are positive constants.

If we observe the population every h days, then the equation expressing the relation between the populations $x(t)$ and $x(t + h)$ at successive observations is

$$\frac{x(t + h) - x(t)}{hx(t)} = R - cx(t).$$

Solve this equation for $x(t + h)$ as the unknown, and put your result in the form

$$x(t + h) = (\quad)x(t) - (\quad)x(t)^2.$$

Fill in the missing coefficients. This equation enables you to predict the population h days from now if you know the population now, at time t.

Before we discuss our theory any further, let us do some numerical experiments.

Exercises
34. Let $R = 0.01, h = 1, c = 0.000\,001$, and $x(0) = 1\,000\,000$. Make a table showing the population at various times.

35. Work out tables for the following cases:

R	h	c	$x(0)$
0.01	1	0.000 001	100 100
0.01	1	0.000 001	90 000
0.01	0.5	0.000 001	1 000 100
0.01	0.01	0.000 001	1 000 100

36. What is the significance of $\lambda = R/c$?

A graphical process

You have obtained an equation of the form

$$x(t + h) = ax(t) - bx(t)^2$$

for predicting the population at time $t + h$ in terms of the population at the time t, with

$$a = (\quad),$$
$$b = (\quad).$$

Fill in the missing values. We can give a graphical process for calculating the prediction. First we draw a graph of the equation

$$y = ax - bx^2.$$

This equation expresses the relation between $x = x(t)$ and $y = x(t + h)$. Do you know what kind of curve represents this relation? If you do not remember, work out the curve in a case with simple numbers like $a = 2$ and $b = 1$. This will surely remind you of this family of curves.

Now if you have the graph of an equation, it is easy to calculate y from x graphically. We shall illustrate this with the case

$$y = F(x) = 4x - \frac{x^3}{3},$$

which is somewhat different from the equation which you should be working on.

The graph of our illustrative equation looks like fig. 6.8. We have also drawn in the line $y = x$. Now if x is given, we locate the corresponding point on the x-axis, go vertically to the curve $y = 4x - (x^3)/3$, then horizontally across to the line $y = x$, then down to the x-axis again. This new point will represent the number $4x - (x^3)/3$. Why?

If you apply this process to the equation

$$y = ax - bx^2$$

and start with $x = x(0)$, the initial population, you will obtain $x(h)$, the population h days later. If you repeat the process, you will obtain successively $x(3h), x(4h)$, etc.

Fig. 6.8

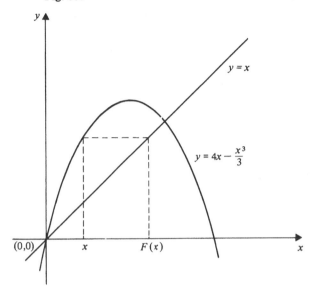

Exercises

37. For what values of $x(t)$ is it true that $x(t + h) = x(t)$? Give the biological and the graphical interpretation.

38. Let E be the non-zero solution of the previous problem. If $x(0) < E$, is $x(h) > x(0)$ or is $x(h) < x(0)$? What happens if you iterate the process? How does $x(t)$ behave for large t?

39. If $x(0) = 1/b$, what is $x(h)$? What happens from then on?

40. If $1/b < x(0) < a/b$, what can you say about $x(h)$? What happens for $t > h$?

41. If $x(0) > a/b$, what can you say about $x(h)$? What is the biological interpretation? Can you suggest any limitation of our model? How might it be improved?

42. Let $z(t) = E - x(t)$ be the deviation of the population from equilibrium at the time t. Show that $z(t + h)$ is related to $z(t)$ by an equation of the form

$$z(t + h) = Az(t) + Bz(t)^2,$$

where A and B are constants. Find formulas for A and B in terms of R, h, and c. Show that if R and c are given then $A > 0$ for all sufficiently small values of h.

43. Show that if $|z(t)| < (1 - A)/B$ then $|z(t + h)| < |z(t)|$.

One species preying on another

Imagine now that we have a lake containing minnows and pike, and that the minnows are part of the food supply for the pike. We assume that these populations are observed every h days, and we denote the populations of minnows and pike at the time t by $x(t)$ and $y(t)$ respectively. As before, we express the laws governing the changes of these populations in terms of the relative rates of change

$$r_x = \frac{x(t+h) - x(t)}{hx(t)} \, , \; r_y = \frac{y(t+h) - y(t)}{hy(t)} \, .$$

Let us examine r_x at a time when the populations are x and y respectively. We assume that there is a certain basic rate of excess of births over deaths for the minnows, given by a positive constant a. There is a correction for the size of the minnow population, which uses up its food supply and poisons the environment, and we assume that this correction is proportional to x. This correction contributes a term $-bx$, where b is a positive constant. Furthermore, the more pike there are, the more they eat the minnows. If we assume a constant rate of consumption of minnows per pike per day, this gives us a correction of the form $-cy$, where c is a positive constant. We thus arrive at the equation

$$r_x = a - bx - cy$$

expressing the relative rate of growth of the minnow population when the minnow and pike populations are x and y respectively.

Reasoning in the same way, we arrive at the equation

$$r_y = A + Bx - Cy,$$

where $A, B,$ and C are positive constants. Notice that the more minnows there are, the more food there is per pike, and the better it is for the pike. This explains the term Bx with a positive coefficient.

We then set up the equations describing how the populations change from the time t to the time $t + h$:

$$\frac{x(t+h) - x(t)}{hx(t)} = a - bx(t) - cy(t),$$

$$\frac{y(t+h) - y(t)}{hy(t)} = A + Bx(t) - Cy(t).$$

Solve these equations for $x(t + h)$ and $y(t + h)$ as unknowns, and express your results in the form

$$x(t+h) = x(t)[(\quad) + (\quad)x(t) + (\quad)y(t)],$$
$$y(t+h) = y(t)[(\quad) + (\quad)x(t) + (\quad)y(t)].$$

Fill in the blanks with the proper coefficients. Note that some of the coefficients are negative.

We can now do some numerical experiments. We can assume numerical values for the coefficients a, b, c, A, B, C, and the time interval h. We can then see what happens if we start out with different initial states $(x(0), y(0))$. We can represent a state of the populations by means of a point (x, y) in the plane. This enables us to picture the various possibilities.

Exercises

44. Assume the following values:

    $a = 0.5, b = 0.000\,001, c = 0.000\,02,$
    $A = 0.01, B = 0.000\,01, C = 0.000\,1.$

    Take $h = 1$. Work out the changes in the populations if the initial populations are $x(0) = 16\,000$ and $y(0) = 2000$. Tabulate the values you find for t, x, and y, up to $t = 25$.

45. In the previous exercise, work out the problem with everything the same except for $h = 0.5$. We say that the state $(x(t), y(t))$ is stationary at $x(t)$ if $x(t + h) = x(t)$. We define states stationary at $y(t)$ similarly.

46. Show on graph paper the set of states (x, y) at which x is stationary in the situation described in exercise 44. Show also, on the same sheet of graph paper, the states at which y is stationary. What is the intersection of these two sets of states?

47. In the situation of exercise 44, what is the set of points (x, y) such that $x > 0, y > 0$? If $x(t) = x$ and $y(t) = y$, then what is the set of points (x, y) such that $x(t + h) > x(t)$? These are the states at which x is increasing. What elementary geometric figure is formed by the points representing these states? What elementary geometric figure is formed by the set of states (x, y) at which y is increasing?

48. Show on your graph paper the sets of points (x, y) at which $x > 0$ and $y > 0$, which represents the following states:
    (a) x and y are both increasing;
    (b) x is increasing and y is decreasing;
    (c) x is decreasing and y is increasing;
    (d) x and y are both decreasing;
    (e) the populations are at equilibrium.

49. Work the above five problems using these values:

    $a = 1, b = 0.1, c = 0.2,$
    $A = -1, B = 0.1, C = 0.1.$

    Take first $h = 1$ and then $h = 0.5$. Take the initial state $x = 2$ and

$y = 3$. If you wish, you may think of x and y as measured in thousands. Notice that now the minnows are the main food supply for the pike, so that if there are not enough minnows, the pike die off. Try also the case where all the values are as above except $A = -1.5$.

50. Set up the general form of the equations describing the situation where two species, say pike and mackerel, prey on the minnows. Try at least one numerical experiment to see what happens if you assume different rates of excess births over deaths and of eating minnows for the two species.
51. Set up the general form of the equations describing the situation where the main food supply of the minnows consists of algae, and the minnows are the main food supply of the pike. Try at least one numerical experiment.
52. The situations as we described them are still far from 'real' situations, and our mathematical models are still too simplistic. To convince you that this is the case, some data on hares and lynxes in Canada are shown in table 6.5.

Table 6.5

Year	Number of hares	Number of lynxes
1882	15 000	30 000
1883	46 000	52 000
1884	55 000	75 000
1885	137 000	80 000
1886	137 000	33 000
1887	95 000	20 000
1888	37 000	13 000
1889	22 000	7 000
1890	50 000	6 000

6.4 Chemical reactions

Reversible chemical reactions also illustrate the concept of equilibrium and the process of approaching it.

For example, if hydrochloric acid (HCl) is poured into an aqueous solution of silver nitrate ($AgNO_3$), a precipitate of silver chloride (AgCl) is formed, since silver chloride is quite insoluble in water. This is explained in terms of a reaction between the silver and chloride ions in the solution:

$$Ag^+ + Cl^- \rightleftarrows AgCl.$$

We have written two-way arrows here, since a small amount of silver chloride is always dissolving in the water and almost immediately dissociates into ions.

The rate at which the reaction takes place from left to right at any given temperature is, to a very good approximation, proportional to the product of the concentrations of the silver and the chloride ions:

$$\text{rate} (\rightarrow) = L [Ag^+] [Cl^-].$$

The constant of proportionality L depends on the temperature. The concentrations are usually measured in moles per liter. The rate at which the reaction takes place in the reverse direction is proportional to the concentration of silver chloride present:

$$\text{rate} (\leftarrow) = R [AgCl].$$

We assume continual stirring.

At any time t, let $x = [AgCl]$, the concentration of silver chloride, let $y = [Ag^+]$, and let $z = [Cl^-]$. We can express the combined effect of the two processes by the equation

$$\frac{dx}{dt} = Lyz - Rx,$$

the first term on the right-hand side representing the rate at which AgCl is formed, and the second representing the rate at which it dissolves and dissociates.

This equation contains three unknowns. To eliminate y and z, we must use the fact that the total amounts present of silver and chlorine do not change. If A is the total concentration of Ag and C the total concentration of Cl, then $x + y = A$ and $x + z = C$. From this we obtain

$$y = A - x, z = C - x, \tag{6.4}$$

and

$$\frac{dx}{dt} = L(A - x)(C - x) - Rx = Q(x). \tag{6.5}$$

These equations must be supplemented by the inequalities

$$x \geqslant 0, \ y \geqslant 0, z \geqslant 0, \tag{6.6}$$

since, obviously, negative amounts are chemically meaningless. Our mathematical model consists of equations (6.4) and (6.5) together with the inequalities (6.6).

Our first step is to analyze the quadratic polynomial Q. From (6.6) and (6.4) we immediately obtain

$$0 \leqslant x \leqslant A \ \text{ and } \ 0 \leqslant x \leqslant C,$$

so that

$$0 \leqslant x \leqslant \min(A, C) = m,$$

where m is the smaller of A and C. We note that

$$Q(0) = LAC > 0 \text{ and } Q(m) = -Rm < 0,$$

so th.t $Q(x)$ has at least one zero in the interval $0 < x < m$. Why?

Also $Q(x) = Lx^2 - [L(A + C) + R]x + LAC$, which is positive for large values of x, so that $Q(x)$ must have another zero greater than m. Since $Q(x)$ cannot have more than two zeros, these are the only zeros of $Q(x)$. Let E be the zero in the interval $0 < x < m$, and r be the other zero, so that

$$0 < E < m < r, Q(E) = Q(r) = 0. \tag{6.7}$$

The factors of $Q(x)$ are then $x - E$ and $x - r$, so that

$$Q(x) = L(x - E)(x - r). \tag{6.8}$$

We can sketch roughly the graph of $Q(x)$ (fig. 6.9). The dashed portions of the

Fig. 6.9

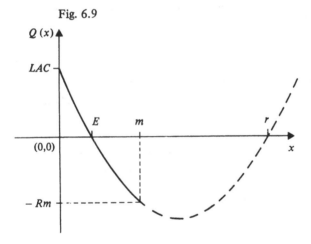

parabola are chemically meaningless. We see that

$$\frac{dx}{dt} = Q(x) > 0 \text{ for } x < E$$

and

$$\frac{dx}{dt} = Q(x) < 0 \text{ for } E < x \leqslant m.$$

Thus x is increasing as a function of t when $x < E$ and x is decreasing as a function of t when $E < x \leqslant m$.

When $x = E$, $Q(x) = 0$, that is, the processes in the two directions *exactly balance* each other. The silver chloride is then dissolving just as fast as it is being formed, so the system is in *equilibrium*. Indeed, the constant $x = E$ is a solution of (6.4), (6.5) and (6.6).

We can obtain a good qualitative idea of what the solutions of (6.4), (6.5) and (6.6) look like by using the method of *direction fields*. On graph paper with t- and x-axes identified we draw through each point (t,x) a small segment with slope $Q(x)$. We only need to consider $t \geqslant 0$ and $0 \leqslant x < m$. We obtain a picture like fig. 6.10, which shows the direction field of the differential equation (6.5). If we draw smooth curves tangent at each point to the direction field we obtain a sketch of the solution curves.

Fig. 6.10

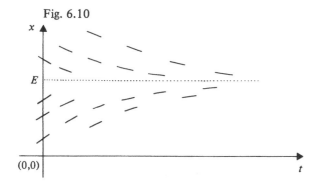

Let $x(0) = x_0$ be the initial amount of silver chloride present. Our picture shows that if $x_0 > E$ then x decreases steadily toward E, whereas if $x_0 < E$ then x increases steadily toward E. Theoretically, according to our model, x never actually reaches E, but the difference $|x - E|$ ultimately becomes so small that we would not be able to detect it with our measuring instruments.

We can easily obtain quantitative results if we try to obtain the *inverse function* $t = t(x)$. Since

$$\frac{dt}{dx} = \frac{1}{dx/dt} = \frac{1}{Q(x)} \text{ ,}$$

we obtain t by integrating the right-hand side. Suppose, for the sake of definiteness $x_0 < E$, so that x is an increasing function of t (fig. 6.11), and

Fig. 6.11

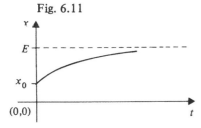

t is an increasing function of x (fig. 6.12). We obtain

$$t = \int_{x_0}^{x} \frac{1}{Q(v)} \, dv. \tag{6.9}$$

Why is the lower limit in the integral equal to x_0?

Fig. 6.12

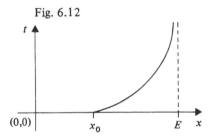

We shall now show how to obtain the most important information about the solution without evaluating this integral.

The integrand has a bad discontinuity at $v = E$, since the denominator has the factor $v - E$ which is zero there. The other factors stay away from zero throughout the interval of interest $0 \leqslant v \leqslant E$. In fact,

$$Q(v) = L(r - v)(E - v) \tag{6.9a}$$

and

$$0 < L(r - E) \leqslant L(r - v) \leqslant Lr \text{ for } 0 \leqslant v \leqslant E,$$

so that

$$L(r - E)(E - v) \leqslant Q(v) \leqslant Lr(E - v)$$

and

$$\frac{1}{Lr} \frac{1}{(E - v)} \leqslant \frac{1}{Q(v)} \leqslant \frac{1}{L(r - E)} \frac{1}{E - v}.$$

Thus we may compare the integral in (6.9) with the much simpler integrals:

$$\frac{1}{Lr} \int_{x_0}^{x} \frac{1}{E - v} \, dv \leqslant t \leqslant \frac{1}{L(r - E)} \int_{x_0}^{x} \frac{1}{E - v} \, dv.$$

If we make the substitutions

$$u = E - v, \, du = -dv,$$

we obtain

$$\int_{x_0}^{x} \frac{1}{E - v} \, dv = \int_{E - x_0}^{E - x} \frac{1}{u}(-du) = \int_{E - x}^{E - x_0} \frac{1}{u} \, du = \log \frac{E - x_0}{E - x}. \tag{6.10}$$

We arrive at the estimate

$$\frac{1}{Lr} \log \left(\frac{E - x_0}{E - x} \right) \leqslant t \leqslant \frac{1}{L(r - E)} \log \left(\frac{E - x_0}{E - x} \right). \tag{6.11}$$

This tells us that t behaves like a constant multiple of the quantity in (6.10). That quantity is large when $E - x$ is small, that is, when x is close to its equilibrium value. For such values of x, the right-hand side of (6.11) is actually a good approximation to t.

It is very rare in practical problems that an integral can be evaluated or a differential equation can be solved explicitly in terms of elementary functions. It is very useful and important therefore, to know how to get good estimates and approximations which give the most essential information about the solutions.

Exercises

53. Try the values $L = 1$, $R = 2$, $A = 2$, $C = 3$ in equation (6.5). Find E and r. If $E - x_0 = 0.5$, estimate from (6.11) how long it will take for $E - x$ to reach the value of 0.05. What about $E - x = 0.005$ or $E - x = 0.0005$?

54. Take $A = 2$, $C = 3$, and let $R/L = \delta$. Find exact formulas for E and r. Use the approximate formula

$$(a + y)^{\frac{1}{2}} = a^{\frac{1}{2}} \left(1 + \frac{y}{2a} \right), \text{ for } |y| \text{ small, } a > 0,$$

to obtain good approximations for E and r when δ is small.

55. In the problem discussed in the text, work out the results analogous to (6.9) and (6.11) if $x > E$. If $L = 1$, $R = 2$, $A = 2$, $C = 3$, and $x_0 - E = 5$, how long will it take for $x - E$ to reach 0.05?

56. Solve the equation

$$Ct = \log \left(\frac{E - x_0}{E - x} \right) \tag{6.12}$$

for x in terms of t. Assign values to x_0, E, and C, and graph x as a function of t. Compare with figs. 6.10 and 6.11. Also graph (6.12) for t as a function of x. Is there a short cut for obtaining one graph from the other? Compare with fig. 6.12.

57. Find constants p and K such that

$$\frac{1}{Q(v)} = \frac{p}{E - v} + \frac{K}{r - v},$$

where Q is as in (6.9a). Find a formula for t in (6.9). Does your result agree with (6.11)? Solve for x in terms of t. Is the result simple? If necessary, look up the method of partial fractions in any standard calculus text.

We have discussed a simple chemical reaction of the type

$$X + Y \rightleftarrows XY$$

with reaction rate laws of the simple form

rate ($\rightarrow$) = constant $[X][Y]$,

rate ($\leftarrow$) = constant $[XY]$.

A somewhat more complicated reaction is that of the decomposition of hydrogen iodide:

$$2HI \rightleftarrows H_2 + I_2.$$

In this case the reaction rates have the form

rate ($\rightarrow$) = $L[HI]^2$

rate ($\leftarrow$) = $R[H_2][I_2]$.

If at any time $t, x = [H_2]$, $y = [I_2]$, $z = [HI]$, our mathematical model would contain the differential equation

$$\frac{dy}{dt} = Lz^2 - Rxy.$$

To express the fact that the total amount of iodine present is constant, we must note that each iodine molecule contains two iodine atoms, so that

$$2y + z = B = \text{constant}.$$

Similarly

$$2x + z = D = \text{constant}.$$

Again we must use the inequalities

$$x \geqslant 0, y \geqslant 0, z \geqslant 0.$$

The mathematical model consists of these three equations together with the inequalities.

Exercises

58. Eliminate x and z and obtain a differential equation involving only the unknown function y. Assign positive values to D, B, L and R and graph the right-hand side as a function of y. Try different assigned values. What difference does it make whether L, and R so-called equilibrium constant) is less than or greater than $\frac{1}{4}$?

59. Analyze this type of reaction in the same way as we analyzed the case of silver chloride.

In many cases it is suspected that the reaction takes place in several steps. It may happen that only the initial amounts and final amounts of certain compounds are accessible to measurement. In such a situation, we may have to make a theoretical analysis of several hypothetical mechanisms with unknown reaction rates and find the functional relation between the measurable quantities. This may help us to decide which hypothetical mechanism fits the observed data best.

By using average rates instead of instantaneous rates, which lead to difference equations rather than differential equations, we can adapt this topic to students who have not studied calculus. Equation (6.5) for AgCl then takes the form

$$\frac{\Delta x}{\Delta t} = Q(x). \tag{6.5a}$$

If we set $\Delta t = h$ and $\Delta x = x(t + h) - x(t)$ and solve for $x(t + h)$, we obtain

$$x(t + h) = x(t) + hQ(x(t)) = P(x(t)), \tag{6.13}$$

where

$$P(x) = x + hQ(x)$$

is a quadratic polynomial. Equation (6.13) is similar to the equation which arose in section 6.3 in the study of the growth of a population. We can treat it in the same way.

The equation

$$s = P(x)$$

describes the transition from the state $x = x(t)$ at time t to the state $s = x(t+h)$ at the time $t + h$. The line $s = x$ describes equilibrium, that is, the situation when the concentration of AgCl remains constant. We can graph these two relations (fig. 6.13). An equilibrium state corresponds to an intersection of the curve $s = P(x)$ with the line $s = x$, that is, a point where

$$Q(x) = 0.$$

Of the two roots E and r of this equation, only E, which satisfies $0 < E < m$, is chemically meaningful.

Now the students can do some numerical experiments.

Exercises

60. With L, R, A, C as in equations (6.4)–(6.6), how does E depend on A for fixed C, or on C for fixed A? Assume values for $L, R,$ and $C,$ solve for E as a function of A, and calculate it for various values of A. Does E increase or decrease as A increases? Similarly, study the dependence of E on C. Is your result valid for arbitrary values of L

Fig. 6.13

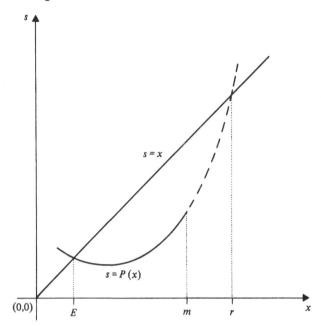

and R? The ratio $R/L = k$ is called the *solubility product*, and is given in tables of physical constants. Look it up. For a given amount of silver nitrate, which determines the value of A, how can you adjust the value of C, which corresponds to the amount of hydrochloric acid, so as to increase the amount E of AgCl formed?

61. Study the dynamics of the problem, that is, how x varies with t. Assume values for L, R, A, C, h, and $x(0)$, and compute $x(t)$ for $t = h$, $2h$, $3h$, ... Does $x(t)$ approach E or not? If h is chosen too large, is the model (6.5a) still realistic? Keeping the ratio $R/L = k$ constant, how does the speed of the approach to equilibrium depend on L? Experimentally, equilibrium is attained very rapidly in this reaction. What does this tell you about the sizes of L and R?

62. Can you give a theoretical explanation of what you found in your numerical experiments in exercises 60 and 61? There may be some advantages in working with the variable $u = x - E$ which describes the deviation from equilibrium.

7

Waves

We felt that a correct name for this chapter would be 'Waves: from kindergarten to graduate work' - but settled for just 'Waves'. The intention remains, though, to treat the notion of waves at as many different levels as we can, while at the same time looking in detail at discrete models of motion and solutions of differential equations.

The two parts of section 7.1 show how the notion of waves can be taught at the kindergarten, the primary level and in the upper elementary school grades. In the first part we use rhythms, patterns, combinations of patterns, and also develop arithmetic skills for the very young. In the second we discuss prime numbers, remainders, periodic patterns, and graphs. The exercises we propose are such that the pupils will make interesting discoveries only if their computations are correct. This entire section is written for the teacher.

Section 7.2, on the vibrating string, consists mainly of a graduated series of exercises with a minimal amount of explanation, and leads the reader from simple deflections to the study of vibrations of an infinite string. This section should be used as a text for the students.

From there we go over to a discrete mathematical model for harmonic motion, various improvements of this model, and discuss how it agrees with the law of conservation of energy. This is not an easy section.

Section 7.4 on trigonometric functions, is squarely written at the college level. In fact we discuss here differential equations, estimates for their solutions, inverse functions - everything but trigonometric functions in their usual form. This unit is also a text for the student.

In this chapter maybe more than in the others we study patterns, their combinations, and how they arise. Therefore this is the place for a word of caution. In mathematics, we know exactly what we mean by a pattern. If we study any phenomenon which seems to obey a certain law, to have a pattern, we will try to guess the law or pattern, but we will not stop there. We will prove (or disprove) our guess. Only at that stage do we establish to our satisfaction that indeed there is a pattern.

Therefore questions like

What is the next number in the series $1, 3, 9, 27, \ldots$?

or in mathematical terms

What is the formula for the nth term of the series $1, 3, 9, 27, \ldots$?

are the wrong *sort* of question. They are misleading because the answers

The next number in this series is 2

or

The formula for the nth term of this given series is
$$2n - 1 + 2(n - 1)(n - 2) + \tfrac{4}{3}(n - 1)(n - 2)(n - 3)$$

are absolutely correct for the questions asked. This example illustrates the point that in mathematics a series (or a pattern) is never determined by giving a few examples or the first few cases. This does not mean that the law must be explicitly stated. Indeed, series or patterns will in most cases be given by recurrence relations, and it will sometimes be very difficult to translate these into an explicit formula. So it is a perfectly good question to ask

In a series every term from the third term on is the sum of the two preceding ones. The first terms are $1, 3, 4, 7, \ldots$ What is the formula for the nth term?

Research problems

1. Let us define the 'general term of a series' as the lowest degree polynomial in n which satisfies the data. Is the question

What is the general term of the series $a_1, a_2, a_3, \ldots$?

legitimate in that case? Would it give the 'expected' answer for the series $2, 4, 8, \ldots$? Can we give a good definition of the term 'expected answer' we just used?

2. What is the maximum number of regions into which n circles divide the plane?

3. Is $n^2 - n + 41$ prime for all n?

7.1 Waves: an elementary approach

Waves in and before the primary grades

One can profitably teach some valuable concepts related to wave motion as early as in kindergarten. The appropriate activities also give, as

by-products, excellent motivated practice in arithmetic skills. You can begin
by calling the children's attention to various repeating patterns which occur
in everyday life, such as

> day, night, day, night, . . . ,
> breakfast, lunch, dinner, breakfast, . . .

You can ask the children to suggest others. The rhythm of familiar tunes also
give repeating patterns, for example, the waltz:

> oom, pah, pah, oom, pah, pah, . . .
> ï,   2,   3,   ï,   2,   3,   . . .

The children can clap or stamp at the 'ooms' while they are counting '1, 2, 3,
1, 2, 3,' etc. They can try march, jig, or reel rhythms as well. If there is a
record of Tchaikovsky's Sixth Symphony available, they might count out the
rhythm of the second movement.

Now you can present this problem: 'Suppose you alternate the desserts
with each of your meals

> apple, orange, apple, orange, . . . ,

and start with an apple for breakfast on Monday. How many meals later will
you again have an apple for breakfast?' They can write two lines on the
blackboard

> B   L   D   B   L   D ...
> A   O   A   O   A   O ... ,

they can make the two patterns with blocks, or the boys can say breakfast,
lunch, dinner, etc.' while the girls say 'apple, orange, apple, orange, etc.'
Thus they can see and hear the new pattern obtained by combining the two
simpler ones.

They can investigate other combinations of repeating patterns in a similar
way. These exercises give the pupils good practice in counting. Later (say,
first grade) they can learn to record their results. There are some interesting
natural questions:

> (a) When you combine two repeating patterns, do you always get a
> repeating pattern?
> (b) If you combine a pattern of 2 with a pattern of 3, do you
> always get a pattern of 6?

They can investigate such questions experimentally.

A little later you can introduce the concept of the *period* of a pattern – the
number of steps before repetition. Then the pupils can make tables of their
observations:

Period of 1st pattern	Period of 2nd pattern	Period of combination
2	3	6

You can raise questions about making predictions.

As the children learn more arithmetic they can notice other repeating patterns:

$0, 0 + 5 = 5, 5 + 5 = 10, 10 + 5 = 15, 15 + 5 = 20, \ldots$
numbers with last digit $0, 5, 0, 5, \ldots$
$10 - 3 = 7, 10 - 7 = 3, 10 - 3 = 7, \ldots$

This last example can be introduced by saying 'Each one pick a number less than 10. Subtract your number from 10. Subtract your answer from 10. Subtract that number from 10. Does anybody notice anything?' This should start a discussion going since each pupil will have a repeating pattern even though they picked different numbers. Ask questions such as:

Is there something special about the number 10?
What would happen if we used some other number?

Another good example is the following rule:

Pick a number less than 10. Now pick another one. These are your first two numbers. Add them and subtract the sum from 30. The answer is your third number. Now add your last two numbers and subtract the sum from 30. The answer is your fourth number. Add the two last numbers again and subtract the sum from 30. This is your fifth number. Go on in the same way. What happens?

Later you can ask whether anyone can make up a similar rule for generating sequences with period 4, or 5.

If you want to give motivated practice in some operation such as multiplication, you can ask each child to make a table such as table 7.1, and to write in the x-column a sequence with period 2, then in the y-column a sequence with period 3. Next they should multiply the numbers in each row and write the answers in the z-column. What happens?

Table 7.1

x	y	$z = xy$
3		
5		
3		
5		
3		
5		

They can experiment with other sequences with the same periods, and with other combinations of periods. They can record their results in a table:

Period of x	Period of y	Period of z
2	3	6

They can also experiment with other operations. You can adjust the numbers and the operation involved so as to give any drill you wish.

Waves in upper elementary school grades

We describe here two activities, suitable for upper elementary school grades, which combine experience with waves and periodic phenomena together with computational practice.

You can begin by writing *in a column* on the board the sequence 1, 2, 4, 8, 16. You then ask 'What is the rule? Copy this sequence and continue until you reach 8192' (giving this number serves as a check for the pupils). You now divide the class into groups, and assign to each group an odd prime number: 3, 5, 7, 11, etc. You ask the children in each group to divide the numbers in the sequence by their assigned prime, and to write the *remainders* in a second column. The pupils in each group can compare their results as a check. Thus the work for the 3-group will look like this:

Sequence	Remainder
1	1
2	2
4	1
8	2
16	1
⋮	⋮

It is almost always necessary at the start to remind the children that when the divisor is larger than the dividend, then the quotient is *zero* and the *remainder equals the dividend.*

The various groups should record their results on the blackboard in a single report (table 7.2). Of course you should ask the pupils what they notice. They usually get quite excited to discover the periodic patterns in the columns.

In the group with the prime 11, the divisions will be laborious for most of the children. (Here calculators cannot be used in any obvious way.) Some of the pupils may not have time to complete the columns for this and the larger primes, which should motivate them to look for short cuts. A hint,

Table 7.2

| | Remainder with prime | | | | |
Sequence	3	5	7	11	13
1	1	1	1		
2	2	2	2		
4	1	4	4		
8	2	3	1		
16	1	1	2		

which may be necessary, is to suggest comparing each remainder with the next one down in any column. Are there any patterns? They will discover that often the remainder is double the previous one. What happens when this simple rule fails? How is the prime at the top of the column involved?

When they discover, discuss, and check the short cut, they can then easily complete the assigned columns. There are now three interesting directions for further investigations:

(*a*) Is there also a periodic pattern for other primes, such as 17, 19, 23, etc.? This can be mini-research for homework. Divide the class into groups, and assign to each group a prime to investigate.

(*b*) What is the relation between the prime and the period? The pupils can make a table:

Prime	Period
3	2
5	4
7	3
.	.
.	.

Is there any pattern or law?

(*c*) Will something similar happen with the powers of other numbers? The students may investigate the powers of 3,

$$1, 3, 9, 27, 81, \ldots ,$$

or 4 or 5, etc. The results for the powers of 10 have an interesting relation to the decimals for the reciprocals of the prime

$$1/3 = 0.33\dot{3}$$
$$1/7 = 0.142\,857\,142\,857\ldots$$

This is an appropriate subject for discussion with children who know about conversion of fractions to decimals.

The pupil can extend the table started in (*b*) above by adding a column for the periods for the powers of each number investigated. This larger table will help in the search for the relation between the primes and the periods. The goal should be to predict, ultimately, the results for cases not yet investigated.

This activity involves many interesting results to be discovered, but the discoveries cannot be made unless one does the computations correctly. Thus it puts a premium on mastery of skills in multiplication and division.

If the children record the numbers in any column on graph paper, they obtain patterns like that shown in fig. 7.1 for the 5-column of table 7.2. Such

Fig. 7.1

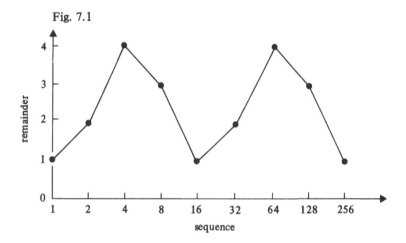

patterns make the wave nature of the phenomenon obvious. The following activity is appropriate for children who know about the multiplication of decimals. Each child is asked to make two columns on his sheet of paper like this

n	x
0	
1	
2	
3	
.	
.	

Each pupil writes any numbers he wishes in the first two rows of the x-column. You may on the blackboard illustrate with numbers of your own. To get the next number in the x-column, multiply the last number by 1.2 and subtract the one before; round off your answers to one decimal place. Thus if you chose 1 and 4, respectively, then your next number would be

$$(1.2 \times 4) - 1 = 4.8 - 1 = 3.8,$$

and you would write this in the 2-row. The next number would be

$$(1.2 \times 3.8) - 4 = 4.56 - 4 = 0.56,$$

and you would write 0.6 in the 3-row. The next number would be

$$(1.2 \times 0.6) - 3.8 = -3.08,$$

and you would write −3.1 in the 4-row. You can continue your own table, explaining as each problem comes up the appropriate rule for computing with signed numbers. Most hand calculators will handle such arithmetic, and if they are available they should be used.

Each child can now compute his own table. The pupils can also record their results on graph paper (fig. 7.2). Each child will get a different wave form, but they will all find that their waves have a period of somewhat more than six units.

Fig. 7.2

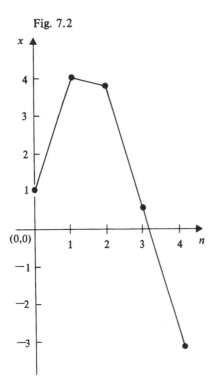

The experiment can be done with any fixed multiplier which is numerically less than 2, instead of 1.2. (With a multiplier greater than 2 one will get more and more violent oscillations.) One can get any period (greater than 1) that one wishes by suitable choice of the multiplier. The initial numbers and the multiplier can be chosen and the number of decimal places used may be chosen so as to give motivated drill in addition, subtraction, and multiplication of decimals, as appropriate to the particular class.

More complicated wave forms can be constructed by calculating another sequence y, using a different multiplier, and then calculating $x + y$. One can also form sums of three or more wave forms. If the computations become too laborious then calculators should be used.

One motivation can be the synthesis of musical tones. You can show photographs of sound waves from various musical instruments. If we wish to imitate the sound of a clarinet, we should first try to analyze the wave of a clarinet tone into a sum of simple waves. We can then use resonators or various electronic devices to produce the desired combination of simple waves. This is the principle of the electronic organ and of the synthesizer. The above activity shows an elementary way to generate simple wave forms, and forming sums shows how to combine them to make more complicated wave forms. It is unfortunately beyond the scope of this book to illustrate the analysis of wave forms into their simple components.

7.2   A vibrating string

We have often observed that whenever we connect mathematics with music, it immediately stirs up interest among the students, even with otherwise indifferent classes. When teaching this unit to children, we usually begin with questions such as:

> Does anyone here play a stringed instrument, a violin or a guitar, for example? Has anyone seen the inside of a piano? Where are the low notes on the piano; the high notes? Which notes come from the short strings; the long strings? How do you play high notes on a violin or guitar?

If possible, it is good to have an instrument available and to illustrate these points.

We then say that if there were time we would perform some experiments first, analyze our observations, and derive a mathematical model from such an analysis. (This would still be the ideal, especially if one had a cooperative and competent colleague in physics.) In lieu of this, the best that we can do is to present a mathematical model without attempting to derive it, and to test its plausibility by comparing the predictions from the model with our qualitative observations.

Suppose now that we want to study how a string moves when stretched tight and clamped at both ends, like a violin string. The best way to make some observations would be to photograph the moving string at equal time intervals, for instance every second. To describe the motion mathematically we divide the string into, say, 12 equal parts, and keep track of the deflections at the endpoints of these parts.

Fig. 7.3 shows schematically how the string looks when it is undisturbed. The endpoints of the parts are labeled 0, 1, ..., 12. The string occupies the

Fig. 7.3

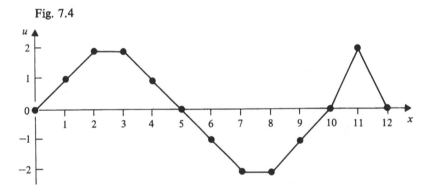

interval, or line segment, $0 \leqslant x \leqslant 12$. The deflection $u(x)$, the deflection at any point x, is zero initially.

If the string is deflected, we approximate the real motion by thinking about $u(x)$ only for $x = 0, 1, 2, ..., 12$. We imagine that the string is straight between any two of these points. When the string is in the position shown in fig. 7.4

Fig. 7.4

the deflection $u(x)$ is given by this table:

x	0	1	2	3	4	5	6	7	8	9	10	11	12
u	0	1	2	2	1	0	−1	−2	−2	−1	0	2	0

Deflections in one direction are counted as positive, and in the opposite direction as negative. The clamping of the string is described by saying that the deflections at the endpoints are zero:

$$u(0) = u(12) = 0.$$

Of course, in reality the string will mostly be a rather smooth curve, and we could obtain a more accurate description by dividing the string into a greater number of parts.

Imagine that we photograph the string every second. We measure the time t by the number of seconds after we start the experiment. We can record the observed deflections in a table 7.3. The clamping of the string is described by writing zero in the 0-column and the 12-column.

Table 7.3

	x =												
t	0	1	2	3	4	5	6	7	8	9	10	11	12
0	0	0	0	0	0	0	0	0	0	0	0	0	0
1	0	1	2	2	1	0	−1	−2	−2	−1	0	2	0
2	0												0
3	0												0
4	0												0
5	0												0

The physical laws of the vibrating string can be described approximately by the rule that in the table, for any four numbers in cells placed like this

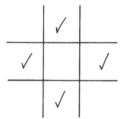

the number at the bottom equals the sum of the two numbers in the previous row minus the number at the top. Thus the number in the 2-row, 3-column, is

	0	
2		1	$2 + 1 - 0 = ?$
	?	

The rule may also be expressed as

> The deflection at any point at any time is the sum of the deflections at the two neighboring points a second before, minus the deflection at the same point two seconds before.

By using this rule, we can predict the shape of the string at the time $t = 2$ from the shape at times 0 and 1, then the shape at time $t = 3$ from the shape at times 1 and 2, and so on.

The preceding elementary analysis can be put in mathematical terms as follows:

The deflection at the point $x = 3$ at the time $t = 0$ (the start) is called $u(3,0)$. The deflection at the point $x = 5$ at the time $t = 3$ (three seconds after the start) is called $u(5,3)$. So the state of the string at the start is described by the numbers $u(0,0), u(1,0), u(2,0), \ldots,$ $u(11,0), u(12,0)$. The state one second later is described by the numbers $u(0,1), u(1,1), \ldots, u(11,1), u(12,1)$.

The physical laws of the vibrating string can be described approximately by the following mathematical equations:

$$u(1,2) = u(0,1) + u(2,1) - u(1,0),$$
$$u(2,2) = u(1,1) + u(3,1) - u(2,0), \text{ etc.}$$

The deflection at the time $t = 2$ at the interior point $x = 4$ is the sum of the deflections at the time $t = 1$ (one second before) at the two neighboring points, $x = 3$ and $x = 5$, minus the deflection at the time $t = 0$ (two seconds before) at the same point $x = 4$.

This law is a good approximation to the actual law of motion. If we divide the string into smaller parts and make our observations every $1/10$ second, we would obtain a more accurate description. You may later learn how to perform the experiments which lead us to this law of motion.

In addition to the above law, we need some *boundary conditions* to tell us what is happening at the ends of the string. We can describe the clamping of the string by the equations:

$$u(0,2) = u(12,2) = 0,$$

and, more generally,

$$u(0,t) = u(12,t) = 0$$

for all t.

Let us do a numerical experiment. We start with our string in an undisturbed state. We pluck it at $x = 3$ at the time $t = 1$ and let go. We show in table 7.4 the states of the string at the times $t = 0$ and $t = 1$. In this table the deflection $u(4,2)$ at the time $t = 2$ and the place $x = 4$ is written in the 2-row, 4-column. According to our law, we must have

$$
\begin{aligned}
u(4,2) &= u(3,1) + u(5,1) - u(4,0) \\
&= 1 \quad\quad + 0 \quad\quad - 0 \\
&= 1.
\end{aligned}
$$

Similarly, we have

$$
\begin{aligned}
u(3,2) &= u(2,1) + u(4,1) - u(3,0) \\
&= 0 \quad\quad + 0 \quad\quad - 0 \\
&= 0,
\end{aligned}
$$

Table 7.4

t	x = 0	1	2	3	4	5	6	7	8	9	10	11	12
0	0	0	0	0	0	0	0	0	0	0	0	0	0
1	0	0	0	1	0	0	0	0	0	0	0	0	0
2	0												0
3	0												0
4	0												0
5	0												0
6	0												0
7	0												0
8	0												0
9	0												0
10	0												0
11	0												0
12	0												0

and

$$u(2,2) = u(1,1) + u(3,1) - u(2,0)$$
$$= 0 \quad + 1 \quad - 0$$
$$= 1.$$

Work out the rest of the state at the time $t = 2$.

To predict the state at the time $t = 3$, we apply the same law. For instance, we have

$$u(3,3) = u(2,2) + u(4,2) - u(3,1)$$
$$= 1 \quad + 1 \quad - 1$$
$$= 1$$

We remind you that the zeros in the 0- and the 12-columns tell us that the string is clamped at both ends.

Exercises

4. Work out the motion of the string with the above states at times $t = 0$ and $t = 1$. You will not see the complete pattern until $t - 24$, but you may be able to predict what happens by the time you reach $t = 13$.
5. Graph the states of the string as in fig. 7.5. You may graph these states (up to $t = 24$) on cards and flip them fast to imitate the moving picture of the vibrating string.

Fig. 7.5

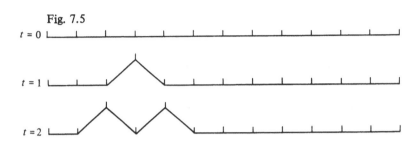

6. Work out the motion for these initial states of the clamped string:

(a)

x =

t	0	1	2	3	4	5	6	7	8	9	10	11	12
0	0	1	0	−1	0	1	0	−1	0	1	0	−1	0
1	0	2	0	−2	0	2	0	−2	0	2	0	−2	0

(b)

x =

t	0	1	2	3	4	5	6	7	8	9	10	11	12
0	0	1	−2	3	0	−3	2	−1	0	1	−2	3	0
1	0	−4	5	6	0	−6	−5	4	0	−4	5	6	0

(c)

x =

t	0	1	2	3	4	5	6	7	8	9	10	11	12
0	0	1	−2	0	2	−1	0	1	−2	0	2	−1	0
1	0	−1	3	0	−3	1	0	−1	3	0	−3	1	0

(d)

x =

t	0	1	2	3	4	5	6	7	8	9	10	11	12
0	0	1	2	−3	4	−2	0	2	−4	3	−2	−1	0
1	0	−1	0	1	−2	1	0	−1	2	−1	0	1	0

7. Try at least three other experiments of choosing initial states for the clamped string and seeing what happens. As in exercises 4 and 5, make 'movies' of the vibrating string for at least two of your experiments.
8. Try experiments with more or fewer subdivisions of the string.
9. Try the simplest situation, with only two subdivisions of the clamped string:

	x =		
t	0	1	2
0	0		0
1	0		0

Try any numbers you like for $u(1,0)$ and $u(1,1)$, and work out the motion. Is the motion periodic? If so, what is the period? Does it make any difference how you choose the initial states?

10. Investigate the next simplest situation, with three subdivisions of the clamped string:

	x =			
t	0	1	2	3
0	0			0
1	0			0

Try any initial states you like, and work out the motion. Is the motion periodic? If so, what is the period? Can you choose the initial states so as to obtain a motion with period 3 seconds? Can you choose the initial states so as to obtain a motion with period 2 seconds?

11. In the same way, investigate four and five subdivisions. Compare. Do you notice anything that can happen with four subdivisions that does not happen with two, three, or five subdivisions? What is the explanation?

12. In exercise 10 work out the motion for any initial states you choose. Let $v(t)$ be the sum of the deflections at the time t:

$v(t) = u(1,t) + u(2,t)$

= the sum of the numbers in the t-row.

Make a table of $v(t)$. What do you notice? Does the same thing happen no matter which initial states you choose? Check that

$v(t + 2) = v(t + 1) - v(t)$

for any time t. Investigate in the same way

$w(t) = u(2,t) - u(1,t)$

= difference of the deflection at $x = 2$
   and the deflection at $x = 1$ at the time t.

What is the general relation which holds between the three numbers $w(t)$, $w(t + 1)$, and $w(t + 2)$?

13. In exercise 11, in the case of four subdivisions, investigate the differences

$a(t) = u(1,t) - u(3,t).$

Is there any general law about the behavior of the sequence $a(t)$? Investigate the behavior of the numbers

$b(t) = u(1,t) + u(3,t)$

and

$c(t) = u(2,t).$

For any motion of the string compute the numbers $b(t+2) + b(t)$ and $c(t+2) + c(t)$, and compare with the numbers $b(t+1)$ and $c(t+1)$. What is the general law?

14. Suppose that the string is 12 units long as in our initial example. Suppose that the end at $x = 12$ is clamped, so that $u(12,t) = 0$ for all $t \geq 0$, but imagine that we wiggle the end at $x = 0$ back and forth. What will happen? Take as initial states:

$x =$												
t 0	1	2	3	4	5	6	7	8	9	10	11	12
0 0	0	0	0	0	0	0	0	0	0	0	0	0
1 1	0	0	0	0	0	0	0	0	0	0	0	0

Our boundary conditions are

$u(0,t) =$ alternately 0 and 1, for $t \geq 0$,

and

$u(12,t) = 0$ for all $t \geq 0$.

In the interior of the string we apply the same law of wave motion as before. Is the motion periodic? If so, what is the period?

15. Try the same experiment as in exercise 14 with other lengths of string. How is the period related to the length?

16. Make 'movies' of your experiments in the last two exercises, as in exercise 5.

17. Try the experiments of exercises 14 and 16 with different motions of the end at $x = 0$. For example, jerk the end every third second:

$u(0,0) = 0, u(0,1) = 0, u(0,2) = 1,$
$u(0,3) = 0, u(0,4) = 0, u(0,5) = 1,$ etc.

How is the period of the motion of the whole string related to the period of the motion of the end at $x = 0$ and the length of the string?

18. Imagine an *infinite* string. We describe its state at the time t by means of its deflections

$\ldots, u(-4,t), u(-3,t), u(-2,t), u(-1,t), u(0,t), u(1,t), \ldots$

at the various points x. Of course now we cannot write down all the deflections at any one time, but we may describe the state by saying how $u(x,t)$ depends on x in general. For example, the initial state

$u(0,x) = 0$ for all x,

is that of an undisturbed string. Plucking the string at $x = 0$ may be described by the equations

$u(x,1) = 0$ for all $x \neq 0$,
$u(0,1) = 1$.

We can describe these two states roughly by the partial table:

$x =$												
t	...	−5	−4	−3	−2	−1	0	1	2	3	4	5 ...
0	...	0	0	0	0	0	0	0	0	0	0	0 ...
1	...	0	0	0	0	0	1	0	0	0	0	0 ...

The symbol ... indicates that the pattern continues.

Work out the motion of this string. Make 'movies' of the motion as in exercise 5. Of course, you cannot now draw the whole string on a card. Draw as much as you have room for. How long does it take the disturbance to reach $x = 10, x = -25$?

19. Try the experiment of exercise 18 with the initial states

$u(x,0) = 0$ for all x,
$u(x,1) = 0$ for $x \neq \pm 1$,

and

$u(1,1) = 1, u(-1,1) = -1$.

Notice how the disturbances produced by this plucking at $x = 1$ and $x = -1$ *interfere* with each other.

20. Try the same experiment with the initial states

$u(0,x) = 0$ for all x,
$u(1,x) = 0$ for $x \neq \pm 1$,

and

$u(1,1) = u(1,-1) = 1$.

Do the disturbances *interfere* with or *reinforce* each other?

21. Try the same experiment with the initial states

$u(x,0) = 0$ for $x \neq 0$,
$u(0,0) = 1$,
$u(x,1) = 0$ for $x \neq 1$,
$u(1,1) = 1$.

Compare the motion with the one resulting from the initial states

$u(x,0) = 0$ for $x \neq 0$, $u(0,0) = 1$,

and

$u(x,1) = 0$ for $x \neq 0$, $u(1,0) = 1$.

22. Try experiments with the same state at $t = 0$ as in exercise 21. Place the disturbance at the time $t = 1$ at various places and see what happens.

23. Try any state you please at time $t = 0$. Consider the state at time $t = 1$

$u(x,1) = u(x - 1,0)$ for all x.

What happens as time goes on?

24. We have used the law

$u(x, t + 2) = u(x - 1, t + 1) + u(x + 1, t + 1) - u(x,t)$

to predict the motion from the states at times $t = 0$ and $t = 1$. As you saw in exercise 18, according to this law a disturbance spreads in both directions at the rate of 1 unit of distance per second. See what happens if you use the law

$u(x, t + 2) = u(x - 2, t + 1) + u(x + 2, t + 1) - u(x,t)$.

Try the experiments of exercises 18 and 21 with this law of motion. Now how does a disturbance spread?

7.3 Simple harmonic motion

The basic equations of Newtonian mechanics

You have probably heard of Newton's laws of motion. In this chapter we shall be concerned with the law of Newton which relates the force acting on a body to its acceleration:

force = mass • acceleration. (7.1)

Let us recall what acceleration is. Suppose that you are driving a car at 40 kilometers per hour and step on the accelerator, and that five seconds later you are moving at 60 kilometers per hour. Your velocity has changed by ( ) kilometers per hour during a time interval of five seconds (or ( ) hours). So your velocity has changed by ( ) kilometers per hour per second, or ( ) kilometers per hour per hour. Fill in the missing values. (Incidentally, how many seconds are there in an hour?) This is your acceleration:

acceleration = rate of change of velocity. (7.2)

Furthermore, you know that

velocity = rate of change of position. (7.3)

These equations (7.1), (7.2), and (7.3) are basic to Newtonian mechanics. If we wish to know how a given mechanical system, say the Solar System or a thrown basketball, will move, we simply write down the above equations for each particle (or body) in the system. If the forces acting on the bodies in the system are known, then we can try to solve the equations for the positions of the bodies at any time.

Exercises

25. Leo dropped a grapefruit from a window of a building. Bruno photographed the experiment and recorded the height x, in meters, of the fruit t seconds after Leo dropped it:

t	x	v (velocity)	a (acceleration)
0	69		
1	62		
2	45		
3	18		

(a) How far did the grapefruit fall during the first second? What was the change in x during the first second? What was the (average) velocity in meters per second, during the first second? (Note that x decreased, so that the change in x was negative. Velocity, the rate of change of x, has direction, in contrast to speed, which does not.) Record this velocity in the v-column.

(b) Calculate the velocity during the second second (i.e., from $t = 1$ to $t = 2$), and record your answer in the next row of the v-column.

(c) What was the change in velocity from $t = 0$ to $t = 1$? What was the acceleration, or change in velocity per second, during this time interval? (Remember that acceleration, like velocity, has direction) Record your answer in the a-column.

(d) Fill in the rest of the above table, as indicated in parts (a), (b), and (c).

(e) What do you notice about the acceleration of the grapefruit?

The equations for simple harmonic motion

We shall now apply Newton's law to the problem of simple harmonic motion. Imagine, for example, a spring with a weight attached to it. We set up an x-axis with the positive direction pointing downwards (fig. 7.6), and choose the origin, $x = 0$, at the equilibrium position.

Fig. 7.6

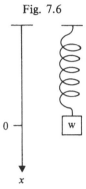

If we displace the weight to the position x (fig. 7.7), then, according to Hooke's law, there will be a restoring force proportional to the displacement. If $x > 0$ (stretching) then the restoring force is (   ) – positive or negative? On the other hand, if $x < 0$ (compressing) the restoring force is (   ) – positive or negative? Fill in the blanks here and in the following text.

Fig. 7.7

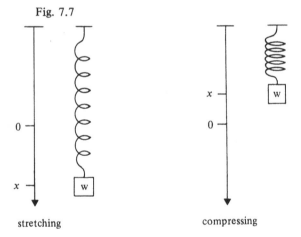

stretching   compressing

Therefore the constant of proportionality must be negative, and Hooke's law takes the form

$$F = -kx, k > 0.$$

Substitute in this equation the value of F in terms of a from the previous equation and solve for a:

$$a = -(\quad)x. \tag{7.4}$$

Imagine that we measure x every h seconds. We observe the position of the spring at

$$t = 0, h, 2h, 3h, \ldots, nh, \ldots$$

and observe the positions

$$x = x(0), x(h), x(2h), x(3h), \ldots, x(nh), \ldots$$

As before, $x(t)$ is the value of x at the time t.

What is the (average) velocity during the time from $t = 0$ to $t = h$? The position changes from $x(0)$ to $x(k)$, so that the change in position is ( ) − ( ). This is the change in x during the first time interval of h seconds. The average velocity during this time interval, which we shall associate with the midpoint of the interval and call $v(h/2)$, is the change in x per unit time:

$$v(h/2) = \frac{\text{change in } x}{\text{change in } t} = (\quad).$$

In the same way, during the next h seconds, x changes from $x(h)$ to ( ). The velocity during this time interval is

$$v(3h/2) = (\quad).$$

More generally, if $v(t + h/2)$ is the velocity in the time interval from t to $t + h$, then

$$v\left(t + \frac{h}{2}\right) = \frac{\text{change in } x}{\text{change in } t} = (\quad). \tag{7.5}$$

Thus corresponding to the times $0, h, 2h$, etc., we get the velocities

$$v = v\left(\frac{h}{2}\right), v\left(h + \frac{h}{2}\right), v\left(2h + \frac{h}{2}\right), \ldots, v\left(nh + \frac{h}{2}\right), \ldots$$

In the same way we can calculate the accelerations during the various time intervals. For example, $a(h)$ is the acceleration in the interval from $h/2$ to $h + h/2$. During these h seconds v changes from $v(h/2)$ to $v(h + h/2)$, so that the change in v is ( ) − ( ). The change in v per unit time is

$$a(h) = \frac{\text{change in } v}{\text{change in } t} = (\quad).$$

More generally, we have

$$a(t) = (\quad).$$

Let us combine this equation with equation (7.4). It is convenient to introduce the quantity $\omega = (k/m)^{\frac{1}{2}}$ so that our equation takes the form

$$\frac{v(t + 3h/2) - v(t + h/2)}{h} = (\quad)\omega^2 x(t + h). \tag{7.6}$$

Solve equation (7.5) for $x(t + h)$ as the unknown, and equation (7.6) for $v(t + 3h/2)$, and make use of equation (7.5):

$$x(t+h) = (\quad)x(t) + (\quad)v\left(t + \frac{h}{2}\right),$$

$$v\left(t + \frac{3h}{2}\right) = (\quad)x(t) + (\quad)v\left(t + \frac{h}{2}\right). \qquad \Bigg\} \ (7.7)$$

What are the coefficients in these equations?

Exercises

26. (a) Let $\omega = 2$, $h = 1$, $x(0) = 3$, $v(1/2) = 0$. Substitute $t = 0$ in equations (7.7) and calculate $x(1)$ and $v(3/2)$.
    (b) Substitute $t = 1$ in equations (7.7) and calculate $x(2)$ and $v(5/2)$.
    (c) Calculate $x(3)$, $v(7/2)$, $x(4)$, and $v(9/2)$.

27. Let $\omega = 2$, $x(0) = 3$, $v(h/2) = 0$. Fill in the following table:

t	$x(t)$	$v(t)$
0		
$h/2$		
h		
$3h/2$		
$2h$		

    (a) For $h = 0.5$ calculate as far as $t = 4$.
    (b) For $h = 0.2$ calculate as far as $t = 2$.
    (c) For $h = 0.1$, here too, go as far as $t = 2$.

28. Compare the values obtained for $x(1)$, $x(2)$, $v(1 + h/2)$, $v(2 + h/2)$ in exercises 26 and 27. What seems to be happening as h becomes smaller?

29. Obtain formulas for $x(t + 2h)$ and $v(t + 5h/2)$ in terms of $x(t)$ and $v(t + h/2)$:

$$x(t + 2h) = (\quad)x(t) + (\quad)v(t + h/2),$$
$$v(t + 5h/2) = (\quad)x(t) + (\quad)v(t + h/2).$$

What are the coefficients?

Since it is inconvenient to work with $v(t + h/2)$, let us change notation and write

$$V(t) = v(t + h/2).$$

We may describe the motion geometrically in the so-called *phase plane*. We take as coordinates V and x, and plot the points (x, V) corresponding to the successive values of t. As we take smaller values of h, the points come closer and closer to a certain curve (fig. 7.8). (Incidentally, physicists usually use position and *momentum* (p = momentum = mV = mass $\times$ velocity) instead of position and velocity, x and V, as the variables in the phase plane.)

Fig. 7.8

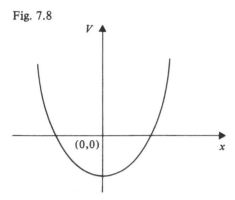

Exercises

30. Graph the results of exercise $27(a)$–(c) in the phase plane.
31. Rewrite the formula obtained in exercise 29 in terms of x and V.

A better choice of variables

As you can see from exercise 31, the formulas for $x(t + nh)$ and $V(t + nh)$ rapidly become more complicated as n increases. The variables x and V are natural to use from the physicist's point of view, but some other variables may be more natural from the mathematical point of view. The mathematician's approach is to look for new variables, instead of x and V, in terms of which equations (7.7) take on a simpler form. Let us see whether we can choose the constant α so that the variable

$$y = x + \alpha V$$

obeys a simpler law. We shall try to choose α so that y obeys the law of a geometric progression:

$$y(t + h) = \lambda y(t),$$

where λ is some constant.

In this equation we substitute

$$y(t + h) = x(t + h) + \alpha V(t + h),$$
$$y(t) = x(t) + \alpha V(t).$$

Then we substitute for $x(t + h)$ and $V(t + h)$ their values from equations (7.7):

$$(\quad + \quad) + \alpha(\quad + \quad) = \lambda[x(t) + \alpha V(t)].$$

Multiply out both sides and write each in the form

$$Ax(t) + BV(t),$$

where A and B are certain coefficients. This equation will be true for all points $x(t)$, $V(t)$ in the phase plane provided that the coefficients of $x(t)$ and $V(t)$ on both sides, respectively, are the same:

$$A = \text{coefficient of } x(t) = 1 - \alpha\omega^2 h = \lambda$$
$$B = \text{coefficient of } V(t) = (\quad) + (\quad). \qquad\qquad \Big\}(7.8)$$

It may help you to see the general idea better if we first solve equations (7.8) for the special case $\omega = 2, h = 0.5$. We have ( ) equations (count them!) and two unknowns, ( ) and ( ). We can solve them by substituting in the second equation the value for λ obtained from the first equation:

$$(\quad) + (\quad) = [(\quad) - (\quad)](\quad).$$

Multiply out the right-hand side. This is an equation for α of degree two:

$$(\quad)\alpha^2 - (\quad)\alpha + 1 = 0.$$

Therefore we find two values for α:

$$\alpha = (\quad) \text{ or } (\quad).$$

The corresponding values of λ are

$$\lambda = (\quad) \text{ or } (\quad).$$

Thus there are two ways in which we can introduce new variables so as to simplify our equations:

$$y = x + \left(\frac{1 + i\sqrt{3}}{4}\right)V, \text{ and } z = x + \left(\frac{1 - i\sqrt{3}}{4}\right)V. \qquad (7.9)$$

In terms of these new variables the equations take on the form

$$y(t + 1) = \left(\frac{1 - i\sqrt{3}}{2}\right) y(t) = \lambda_1 y(t),$$
$$z(t + 1) = \left(\frac{1 + i\sqrt{3}}{2}\right) z(t) = \lambda_2 z(t). \qquad\qquad \Big\}(7.10)$$

These are just like the equations for our models in the sections on the struggle for life and radioactive decay in chapter 3. We obtain

$$y(t) = (\quad)^? y(0),$$

and

$$z(t) = (\quad)^? z(0).$$

To obtain the description of the motion in terms of our original variables x and V, we solve equations (7.9) for x and V:

$$x = (\quad)y + (\quad)z,$$
$$V = (\quad)y + (\quad)z.$$

We thus express $x(t)$ and $V(t)$ in terms of $y(t)$ and $z(t)$, then $z(t)$ in terms of $y(0)$ and $z(0)$, and these, in turn, in terms of $x(0)$ and $V(0)$:

$$x(t) = \beta(t)x(0) + \gamma(t)V(0),$$
$$v(t) = \delta(t)x(0) + \epsilon(t)V(0),$$

$$\Big\} (7.11)$$

where the coefficients are:

$$\beta(t) = (\quad),$$
$$\gamma(t) = (\quad),$$
$$\delta(t) = (\quad),$$
$$\epsilon(t) = (\quad).$$

Exercises

32. Find $x(t)$ and $V(t)$ if $x(0) = 1$, $V(0) = 0$ (keeping $\omega = 2, h = 0.5$). Do the formulas you obtain from (7.11) look as though they give real numbers for $x(t)$ and $V(t)$ when $t = 2, 3,$ and 4? Check and see whether they do.

33. Calculate $\delta(3)$ and $\epsilon(4)$ (still with $\omega = 2, h = 0.5$).

34. Define $C(t)$ by

$$C(t) = |y(t)|^2,$$

(remember that $y(t)$ is a complex number) and let $\omega = 2, h = 0.5$. Compare $C(t + 1)$ with $C(t)$. What do you observe?

35. Now take again $\omega = 2$, but let h be any positive number. Solve equations (7.8) in this case and obtain formulas for α and λ in terms of h. (Remember to keep track of which value of λ goes with which value of α.) Define $C(t)$ as in exercise 34 and show that

$$C(t + h) = \eta C(t)$$

where η is a certain constant. Does η depend on h?

36. What curve is formed by the set of points (x,V) in the phase plane for which

$$\left(x + \frac{V}{4}\right)^2 + \frac{3V^2}{4} = C,$$

where C is a constant? What family of curves is represented by this equation for different values of C?

37. (a) Solve equations (7.8) for α and λ when ω and h are any positive numbers. Find formulas for $x(t)$ and $V(t)$ in terms of $x(0)$ and $V(0)$.

(b) Let

$$C(t) = |y(t)|^2.$$

Prove that

$$C(t) = K^t C(0)$$

where K depends on ω and h.

(c) If $h\omega$ is sufficiently small, does K depend on h? What is the critical value of $h\omega$?

38. (Uses de Moivre's theorem.) In exercise 37 obtain the formula for $y(t)$ in terms of $y(0)$ and express in trigonometric form. What about $z(t)$?

Conservation of energy

In exercise 37 above, you found that the quantity

$$[x(t) + \frac{h}{2} V(t)]^2 + \left(\frac{4 - h^2\omega^2}{4\omega^2}\right) V^2(t)$$

is a constant for all t. If you use the relation $\omega^2 = k/m$, write V for $V(t)$, and multiply by the constant $k/2$, you can put this result in the form

$$\frac{k}{2}\left[x + \left(\frac{h}{2}\right)V\right]^2 + \frac{m}{2}\left(1 - \frac{h^2\omega^2}{4}\right)V^2 = E, \tag{7.12}$$

which is constant throughout any motion. Thus this function of the phase (x, V) does not change during the motion. Such a law for a physical system is called a *conservation* law.

In the mathematical model we have constructed in order to describe this mechanical system, we imagine that we measure the position x every h seconds. However, the spring is moving *continuously*; our model is therefore an approximation, and the approximation would be better if h were smaller. This suggests that to find out what really happens we should let h *approach* zero, but as h approaches zero, V approaches v, $x + hV/2$ approaches x, and $[1 - (h^2\omega^2)/4]$ approaches 1. This suggests that the *true* conservation law is

$$kx^2/2 + mv^2/2 = \text{constant} = E \tag{7.12a}$$

throughout any motion.

This is the fundamental *law of the conservation of energy*. The terms $kx^2/2$ and $mv^2/2$ are called the *potential* and *kinetic* energies respectively, and their sum E is called the total energy. The law says that during any motion, the total energy of the system remains constant although there may be transfer, or conversion, from one form of energy to another.

These considerations suggest that in our model we ought again to interpret E as the total energy, and the terms $k(x + hV/2)^2/2$ and $mV^2(1 - h^2\omega^2/4)/2$ as the potential and kinetic energies, respectively. Then the law of conservation of energy still holds in our discrete time model in this modified form.

We can interpret the conservation law geometrically in terms of the graph of (7.12) in the phase plane (fig. 7.9). This graph (if $0 \leqslant h\omega < 2$) is an ellipse, with center at the origin, whose axes are rotated through a small

Fig. 7.9

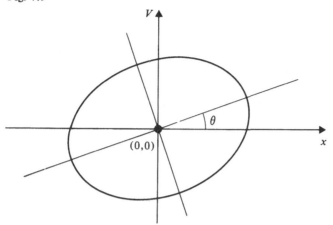

angle θ. For different values of E we get a family of similar and similarly-placed ellipses. As h approaches zero, θ also approaches zero. For a given initial point $(x(0), V(0))$, E can be calculated from (7.12). As time goes on the point (x, V) in the phase plane, which represents the state of the system moves around the ellipse.

Exercises

39. (a) Graph the motions you tabulated in exercise 27(a)-(c). How does the point move around the ellipse, clockwise or counter-clockwise?

    (b) Using the same values for h and ω, try various initial points in the four quadrants. Calculate a few more points, graph them, and say how the ellipse is traversed.

40. Suppose we associate the velocities and accelerations with the *initial* instant of each time interval. Then the equations take on the apparently simpler form

$$\frac{x(t+h) - x(t)}{h} = v(t),$$

$$\frac{v(t+h) - v(t)}{h} = -\omega^2 x(t).$$

Derive equations of the form

$$x(t+h) = (\quad)x(t) + (\quad)v(t),$$
$$v(t+h) = (\quad)x(t) + (\quad)v(t),$$

to describe the transition of the state at time t to the state at time

$t + h$. Calculate α and λ for this new system of equations. Let $C(t) = |y(t)|^2$. How does $C(t + h)$ compare with $C(t)$? What happens to the energy in this model? Calculate a few motions with the same values of ω and h as you used before. What do the paths in the phase plane look like now? Do you still get the same *limiting* law as h approaches zero?

41. State the law of conservation of energy in terms of the variables x and p ($p = mv$).

42. Suppose there is a fractional force proportional to the velocity:

$$F = -kx - cv,$$

where c is a positive constant. Set up equations, introduce the new variables y and z as before, and find formulas for the two values of α and λ. Prove that if h is sufficiently small then $|\lambda| < 1$. What happens to $|y(t)|$ and $|z(t)|$ as t grows larger and larger? What does this imply about the motion of the spring?

Remarks on teaching simple harmonic motion

Simple harmonic motion is usually taught using calculus, as in section 7.4. In this section we have used average, rather than instantaneous, velocities and accelerations, which lead us to the difference equations (7.5) and (7.6). These are approximations to the differential equations

$$\left. \begin{array}{l} x' = v, \\ v' = -\omega^2 x, \end{array} \right\} (7.13)$$

which we will discuss in section 7.4.

As we see, the difference equations can be treated using only high school mathematics. In the course of this treatment we encounter, in a natural way, the topics of

> linear transformations,
> simultaneous equations,
> quadratic equations,
> complex numbers,
> the equation of an ellipse, and
> de Moivre's theorem,

all of which occur in the high school curriculum.

The solutions of (7.13) with the initial conditions $x(0) = 0$, $v(0) = 1$, are

$$x = (1/\omega)\sin(\omega t), \quad v = \cos(\omega t).$$

Equations (7.7) are easy to program. If the students compute $x(t)$ and $v(t + h/2)$ for $\omega = 1$, h small, and $x(0) = 0$, $v(h/2) = 1$, they will obtain good approximations to the trigonometric functions, which they can compare with the standard tables. Actually, when you use a command like

LET A = SIN(B)

in a program, the computer uses a built-in program similar to the one for equations (7.7). To put a table of sines into the computer would be an unnecessary burden on its memory.

7.4 Trigonometric functions

Differential equations

The reasoning of section 7.3, formulated in terms of instantaneous velocities and accelerations, leads to *differential* equations instead of *difference* equations. Now we have

$$\frac{dx}{dt} = v$$

and

$$F = -kx = ma = m\frac{dv}{dt},$$

which yields the system

$$\left.\begin{array}{l} \dfrac{dx}{dt} = v \\[2mm] \dfrac{dv}{dt} = -\omega^2 x, \end{array}\right\} \quad (7.14)$$

where $\omega = (k/m)^{\frac{1}{2}}$, of two equations with two unknown functions.

This system resembles the single equation for the exponential function, but there are some important differences. As before, let us see what we can find out about the solutions directly from the equations.

The difference equations of section 7.3 can be regarded as approximations to (7.14), and can be used with small values of h to compute the approximate solutions. This is easy to program for a computer, and it is one practical way to calculate the solutions of (7.14).

Let us first approach equations (7.14) qualitatively. We can regard a solution $x = x(t)$, $v = v(t)$ as a curve in the phase plane x-v. The tangent to this curve at $t = 0$ is the line

$$x = x(0) + x'(0)\lambda, \ v = v(0) + v'(0)\lambda, \ -\infty < \lambda < +\infty$$

through the point $(x(0), v(0))$ in the direction of the vector $(x'(0), v'(0))$. If we calculate these derivatives from (7.14), we find

$$(x'(0), v'(0)) = (v(0), -\omega^2 x(0)).$$

Similarly, at any point $(x(t_0), v(t_0))$ of the solution curve the tangent is in the direction of the vector $(v(t_0), -\omega^2 x(t_0))$.

Fig. 7.10

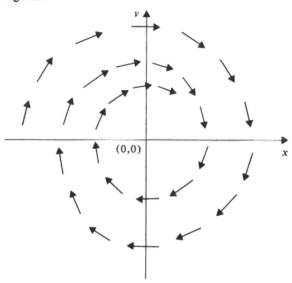

(0,0)

Let us draw, at several points (x,y) of the phase plane, a small directed line segment in the direction $(v, -\omega^2 x)$. This gives us a sketch of the so-called *direction field* of the equation (7.14); fig. 7.10 is a sketch of the direction field with $\omega = 2$. Any solution curve of (7.14) must be tangent to this direction field at any point. Sketch a few solution curves. Clearly the solution curves must wind around the origin. Unless we make our sketches very carefully, it is hard to tell whether we get closed curves or spirals around the origin. In most practical problems it is not worth while to put in the effort of making a very accurate drawing to decide such delicate questions.

Exercises

43. Suppose you set $t = c\tau$, $v = by$, where c and b are constants. What form do equations (7.14) have in terms of the variables τ, x, and y? Can b and c be chosen so as to put the equations in a simpler form? (This amounts to changing the units of time and velocity.)

44. Sketch the direction field for $\omega = 1$. Sketch a few solution curves.

45. Suppose $x = x(t)$ and $v = v(t)$ are solutions of (7.14). Let $X = Ax(t)$, $V = Av(t)$, where A is any constant. Compute

$$\frac{dX}{dt} - V, \frac{dV}{dt} + \omega^2 X.$$

46. Let $(x_1(t), v_1(t))$ and $(x_2(t), v_2(t))$ both be solutions of (7.14). Let $X = x_1(t) + x_2(t)$, $V = v_1(t) + v_2(t)$, and compute

$$\frac{dX}{dt} - V, \frac{dV}{dt} + \omega^2 X.$$

47. Let $(x(t), v(t))$ be a solution of (7.14). Let $X(t) = x(t+3)$, $V(t) = v(t+3)$, and compute $X'(t) - V(t)$, $V'(t) + \omega^2 X(t)$. Can you generalize the result?

48. Let $(x(t), v(t))$ be a solution of (7.14), and let $z = Bx^2 + v^2$, where B is any constant. Compute

$$\frac{dz}{dt}.$$

Which choice of B gives the simplest result?

Estimates of the solution

Let $(x(t), v(t))$ be a solution of (7.14), and suppose that $\omega = 1$. From exercise 48,

$$C = x^2 + v^2 \tag{7.15}$$

is a constant. Hence the solution curve is a circle with radius $\sqrt{C}$. This is the mathematical expression of the law of conservation of energy for the spring.

Suppose $x(0) = 0$, $v(0) = 1$. Then

$$x^2 + v^2 = 1 \tag{7.16}$$

for all t, so that

$$|x| \leqslant 1, |v| \leqslant 1 \tag{7.17}$$

for all t.

Starting with (7.17), we can obtain successively better estimates for the solution. For example, from (7.14) and (7.17), we have

$$x' = v \leqslant 1.$$

As 1 is the derivative of t we see that

$$(x - t)' = x' - 1 \leqslant 0,$$

from which it follows that

$$x - t \text{ is non-increasing.}$$

In particular, for $t \geqslant 0$, we have

$$x(t) - t \leqslant x(0) - 0 = 0,$$

so that

$$x \leqslant t \text{ for } t \geqslant 0, \tag{7.18}$$

Substituting this in the other equation of (7.14), we find

$$v' = -x \geqslant -t \text{ for } t \geqslant 0.$$

(What function has the derivative t?). Hence we infer

$$(v + t^2/2)' \geqslant 0,$$

and

$v + t^2/2$ is non-decreasing.

This yields

$$v(t) + t^2/2 \geqslant v(0) + 0^2 = 1 \text{ for } t \geqslant 0.$$

We can combine this with (7.17) to obtain the sandwich

$$1 - t^2/2 \leqslant v \leqslant 1 \text{ for } t \geqslant 0. \tag{7.19}$$

We can continue this process, step by step. Combining (7.19) with (7.14),

$$x' = v \geqslant 1 - t^2/2 \text{ for } t \geqslant 0,$$

we now look for a function whose derivative is $1 - t^2/2$. This yields:

$$x - t + t^3/6 \text{ is non-decreasing for } t \geqslant 0,$$

so that

$$x(t) - t + t^3/6 \geqslant x(0) - 0 + 0^3/6 = 0,$$

and

$$t - t^3/6 \leqslant x.$$

This, together with (7.18), gives the sandwich

$$t - t^3/6 \leqslant x \leqslant t \text{ for } t \geqslant 0. \tag{7.20}$$

Already (7.20) enables us to compute x quite accurately for small values of t. If we use this sandwich and $v' = -x$, the above reasoning now gives us

$$1 - t^2/2 \leqslant v \leqslant 1 - t^2/2 + t^4/24 \text{ for } t \geqslant 0. \tag{7.21}$$

This enables us to compute v quite accurately for small values of t. If we continue with the same method, we can obtain formulas which give us even better approximations for x and v.

Exercises

49. If $x(0) = v(0) = 0$, what is C in (7.15)? What are x and v for any value of t?

50. Suppose $(x_1(t), v_1(t))$ is a solution of (7.14) such that $x_1(0) = x(0) = v$, $v_1(0) = v(0) = 1$. Let $X(t) = x_1(t) - x(t)$, $V(t) = v_1(t) - v(t)$. Use exercises 45, 46, and 49 to find out what X and V are for any value of t. What does this tell you about $x_1(t)$ and $v_1(t)$?

51. If $(x(t), v(t))$ is a solution of (7.14), let $X = v$, $V = -x$, and compute $X' - V$ and $V' + X$.

52. If (x,v) is the solution of (7.14) such that $x(0) = 0$, $v(0) = 1$, express in terms of x and v the solution $(x_1(t), v_1(t))$ such that $x_1(0) = 1$, $v_1(0) = 0$.

53. If (x,v) and (x_1,v_1) are the solutions of (7.14) mentioned in exercise 52, let
$$X = Ax + Bx_1, \quad V = Av + Bv_1.$$
Determine the constants A and B so that
$$X(0) = 2, \quad V(0) = 3.$$
What do exercises 45, 46 and 50 tell you about (X, V)?

54. Find the smallest positive solution t_0 of the equation
$$1 - t^2/2 + t^4/24 = 0.$$
What can you say about $v(t_0)$, where (x,v) is the solution of (7.14) such that
$$x(0) = 0, v(0) = 1?$$

55. If (x,v) is the solution of (7.14) such that $x(0) = 0$, $v(0) = 1$, can $v(t)$ be positive in the entire interval $0 \leqslant t \leqslant \sqrt{6}$? How many solutions does the equation $v(t) = 0$ have in this interval? For these values of t, evaluate $x(t)$.

56. Suppose (x,v) is the solution of (7.14) such that $x(0) = 0$, $v(0) = 1$, and that c is a number such that $x(c) = 1$, $v(c) = 0$.
   (a) Use the principles of exercises 47 and 53 to find simple formulas for $x(t + c)$ and $v(t + c)$ in terms of $x(t)$ and $v(t)$.
   (b) Find formulas for $x(t + 2c)$, $v(t + 2c)$, $x(t + 4c)$, $v(t + 4c)$ in terms of $x(t)$ and $v(t)$.

57. Find a sandwich for the smallest positive c in exercise 56.

58. Obtain an improved sandwich on x from (7.21) above.

59. Obtain a still better sandwich on v.

60. Predict the results of applying the reasoning of the text to improve the sandwiches in exercises 58 and 59. Check your predictions.

61. In which time interval do the two sides of the sandwich for v in (7.21) agree to within an error of less than $0.000\,05 = 5 \times 10^{-5}$? Find similar intervals for the sandwiches in (7.20) and exercises 58 and 59. If (x,v) is the solution of (7.14) such that $x(0) = 0$, $v(0) = 1$, let $X(t) = -x(-t)$, $V(t) = v(-t)$. Compute $X' - V$ and $V' + X$. Apply the results of exercise 50.

Separation of variables: the inverse functions

Another approach to the study of the solution of (7.14) such that $x(0) = 0$, $v(0) = 1$ is to apply the method of separation of variables. If we solve equation (7.16) for v in terms of x (using the fact that, by (7.21), $v \geqslant 0$ for $0 \leqslant t \leqslant \sqrt{2}$), we obtain
$$v = (1 - x^2)^{\frac{1}{2}} \quad \text{for } 0 \leqslant t \leqslant \sqrt{2}.$$

We put this information into equation (7.14):

$$\frac{dx}{dt} = (1 - x^2)^{\frac{1}{2}}.$$

As we did when we studied the exponential function, it is convenient to work with t as a function of x, that is, the *inverse* function of $x(t)$. We obtain

$$\frac{dt}{dx} = \frac{1}{dx/dt} = \frac{1}{(1 - x^2)^{\frac{1}{2}}} ;$$

in other words, dt/dx is this known function of x. Hence we find that

$$t = \int_0^x \frac{1}{(1 - \lambda^2)^{\frac{1}{2}}} \, d\lambda. \tag{7.22}$$

The various methods which we used for evaluating or estimating integrals can now be applied. The integrand

$$g(\lambda) = 1/(1 - \lambda^2)^{\frac{1}{2}}$$

is continuous for $0 \leqslant \lambda < 1$. It has a bad discontinuity at $\lambda = 1$, and its value is imaginary for $\lambda > 1$.

We can use equation (7.22) to calculate t for $0 \leqslant x < 1$. There are two devices which we can use to improve the calculation or estimation of the integral. One approach is to notice that

$$1 - \lambda^2 = (1 - \lambda)(1 + \lambda);$$

$1/(1 + \lambda)^{\frac{1}{2}}$ is a nice continuous function over the whole of the interval $0 \leqslant \lambda \leqslant 1$, so that the trouble at $\lambda = 1$ comes entirely from the factor $1 - \lambda$. We can immediately obtain the crude estimates

$$1 \leqslant (1 + \lambda)^{\frac{1}{2}} \leqslant \sqrt{2} \text{ for } 0 \leqslant \lambda \leqslant 1$$

and

$$\frac{1}{\sqrt{2}(1 - \lambda)^{\frac{1}{2}}} \leqslant \frac{1}{(1 - \lambda^2)^{\frac{1}{2}}} \leqslant \frac{1}{(1 - \lambda)^{\frac{1}{2}}},$$

yielding

$$\frac{1}{\sqrt{2}} \int_0^x \frac{d\lambda}{(1 - \lambda)^{\frac{1}{2}}} \leqslant t \leqslant \int_0^x \frac{d\lambda}{(1 - \lambda)^{\frac{1}{2}}}.$$

The integral on the right can be evaluated by means of the substitution $u = 1 - \lambda$, $du = -d\lambda$:

$$\int_0^x (1 - \lambda)^{-\frac{1}{2}} \, d\lambda = \int_{1-x}^1 u^{-\frac{1}{2}} \, du$$

$$= [2u^{\frac{1}{2}}]_{(1-x)}^1$$

$$= 2[1 - (1 - x)^{\frac{1}{2}}].$$

Thus we obtain the sandwich

$$\sqrt{2}[1-(1-x)^{\frac{1}{2}}] \leqslant t \leqslant 2[1-(1-x)^{\frac{1}{2}}]. \tag{7.23}$$

We can improve this result if we subdivide the interval $0 \leqslant \lambda \leqslant x$ into several subintervals, and apply the above method to each of them. For example, we could start with

$$\int_0^x \frac{d\lambda}{(1-\lambda^2)^{\frac{1}{2}}} = \int_0^{x/3} \frac{d\lambda}{(1-\lambda^2)^{\frac{1}{2}}} + \int_{x/3}^{2x/3} \frac{d\lambda}{(1-\lambda^2)^{\frac{1}{2}}} + \int_{2x/3}^x \frac{d\lambda}{(1-\lambda^2)^{\frac{1}{2}}}.$$

If we use a fine subdivision, we shall get greater accuracy. One interesting conclusion from (7.23) is that the time t_0 when x reaches 1 (and ν reaches 0) is between $\sqrt{2}$ and 2. If we divide the interval $0 \leqslant \lambda \leqslant 1$ into many small subintervals, we can estimate t_0 as accurately as we wish.

A second approach is to approximate the integrand by polynomials. We shall illustrate this with a related, but simpler, integral. This integral arises when we consider the ratio

$$w = x/\nu$$

and the differential equation it satisfies. We find that if $x(0) = 0$, $\nu(0) = 1$, then

$$w' = \frac{\nu x' - x\nu'}{\nu^2} = \frac{\nu^2 + x^2}{\nu^2} = 1 + w^2,$$

and

$$w(0) = \frac{x(0)}{\nu(0)} = \frac{0}{1} = 0.$$

Hence the inverse function satisfies

$$\frac{dt}{dw} = \frac{1}{1+w^2},$$

and

$$t = \int_0^w \frac{1}{1+\lambda^2} d\lambda. \tag{7.24}$$

If we divide $1 + \lambda^2$ into 1 by the usual algorithm, we obtain

$$\left.\begin{aligned}
\frac{1}{1+\lambda^2} &= 1 - \lambda^2 + \frac{\lambda^4}{1+\lambda^2} \\
&= 1 - \lambda^2 + \lambda^4 - \frac{\lambda^6}{1+\lambda^2} \text{, etc.}
\end{aligned}\right\} \tag{7.25}$$

From these, we obtain

$$1 - \lambda^2 \leqslant \frac{1}{1+\lambda^2} \leqslant 1 - \lambda^2 + \lambda^4.$$

If we use this in (7.24), we get the sandwich

$$w - \frac{w^3}{3} \leqslant t \leqslant w - \frac{w^3}{3} + \frac{w^5}{5}. \tag{7.26}$$

If we were to use more terms in (7.25), we would obtain different sandwiches. If w is sufficiently small, estimates like (7.26) give us very good approximations to t.

Exercises

62. Calculate t from (7.22) for $x = 0, 0.1, 0.2, 0.3, \ldots, 0.9, 1.0$, using the first approach above.

63. For which values of w do the estimates in (7.26) differ by less than 5×10^{-5}, so that you get t correct to four decimal places?

64. If you use one more term in (7.25) to get an estimate like (7.26), for which values of w will you obtain t correct to six decimal places? Will your new sandwich for t be an improvement for $w = 1.1$?

65. Let $y = (1 - x)^{-\frac{1}{2}}$. Find a constant c such that

$$(1 - x)y' = cy. \tag{7.27}$$

Try substituting

$$y = 1 + ax + bx^2 \tag{7.28}$$

in (7.27), where a and b are constants. Can a and b be chosen so that this polynomial (7.28) satisfies (7.27) exactly? Can they be chosen so that the coefficients of x^0 and x^1 on both sides of (7.27) agree?

66. Let $Y = 1 + ax + bx^2$ be the polynomial you obtained in the last part of exercise 65 and let $y = (1 - x)^{-\frac{1}{2}}$. Calculate

$$(1 - x)\left(\frac{Y}{y}\right)'.$$

Is $(Y/y)'$ positive or negative for $0 \leqslant x < 1$? Is Y/y greater or less than 1 for $0 < x < 1$? Find a constant $k > 0$ such that

$$-\left(\frac{Y}{y}\right)' \leqslant ky$$

for $0 \leqslant x \leqslant 0.5$. Find an upper estimate for $1 - Y/y$ for $0 \leqslant x \leqslant 0.5$.

67. Use the results of exercise 66 to get a good approximation to the integrand of (7.22) by a fourth degree polynomial in λ for sufficiently small λ. Obtain an approximate formula for t. Can you estimate the error? Compare with your computation in exercise 62.

68. Estimate the time t_0 when $x(t_0) = 1$ from exercise 62. How does this compare with the estimate obtained in the preceding text?

69. Use the results of exercise 62 to graph the relation $x = x(t)$ for

$0 \leqslant t \leqslant t_0$. Use $v = (1 - x^2)^{\frac{1}{2}}$ to graph $v = v(t)$ for $0 \leqslant t \leqslant t_0$.

70. Use the results of exercises 56 and 69 to graph $x = x(t)$ and $v = v(t)$ for $0 \leqslant t \leqslant 4t_0$. What does the continuation for larger values of t look like?

71. Use the method of exercises 65 and 66 to obtain a third degree polynomial which almost satisfies (7.27). Estimate the error as in exercise 66. Use this to improve the result of exercise 67.

8

Linear algebra

Abstract algebra is not generally taught by examining its applications. Indeed, there is even a trend towards avoiding the intuitive and the applied and insisting on the abstract and axiomatic points of view. This might be right for mathematics majors, but is certainly wrong for future teachers. We propose an application-oriented approach in this chapter.

In section 8.1, a step by step, very detailed discussion of changes of temperature in a thin rod leads us to four-dimensional vectors, linear transformations, bases, eigenvalues and eigenvectors. The exercises are an integral part of the exposition and should be done very carefully.

The next section is written as a text at the eighth grade level. We stray here somewhat from our usual procedure, which is to introduce the mathematical notions as they develop while we examine the application. Instead, the application has been squarely based on the use of matrices. Still, it is an effective way of teaching matrices since cryptography is such an attractive subject. Matrix multiplication, inverses, and arithmetic modulo 26 are discussed at a basic level.

The natural algebra of linear differential operators is studied in section 8.3, up to the mathematical formulation of Heisenberg's uncertainty relation and the solution of some linear differential equations. Remarks for the teacher on how to introduce the algebraic aspects of calculus end this section.

8.1   Heat conduction II

The difference equation
We shall start with heat conduction in a thin insulated rod. We locate a point on the rod by its distance x from the left endpoint. The endpoints are labeled 0 and L, where L is the length of the rod. The temperature in degrees centigrade, at the time t at the point x, is denoted by $u(t,x)$.

Let us consider a simple experiment. We start out with the rod at the (room) temperature of 20 °C. At the beginning of the experiment we plunge the left endpoint ($x = 0$) into ice-water and the right endpoint ($x = L$) into a jet of steam. From then on, ($t > 0$), we keep these endpoints at the temperatures 0 °C and 100 °C respectively. How will the temperature on the rod change with time?

We can state the problem in mathematical language. The *initial condition*, that is, the temperature distribution at the start, is that

$$u(0, x) = 20 \text{ for } 0 < x < L.$$

The *boundary conditions*, that is, the situations at the endpoints, are that

$$u(t, 0) = 0 \text{ for } t \geq 0,$$
$$u(t, L) = 100 \text{ for } t \geq 0.$$

We still have to state in mathematical language how the temperature at various times and places are related. Then we shall be ready to predict the value of $u(t, x)$ for any $t > 0$ and any x between 0 and L.

As a first approximation, let us imagine that we observe the temperatures only every h minutes, and we record our measurements at the times $t = 0$, $h, 2h, 3h, 4h, \ldots, nh, \ldots$ We measure the temperatures only at certain points. Suppose also that the rod is 5 cm long ($L = 5$), and that we observe the temperatures at the points $x = 0, 1, 2, 3, 4$, and 5 (fig. 8.1). We can record

Fig. 8.1

our observations in tabular form; in table 8.1, for example, we have taken $h = 1$. At the time $t = 1$, the temperatures at the endpoints of the segment from points 3 to 4 are $u(1,3)$ and $u(1,4)$. As a model of the given situation we could then describe the flow of heat energy from the point $x = 4$ into the point $x = 3$ by

$$k[u(1,4) - u(1,3)],$$

where k is a constant of proportionality. Of course, we count a flow *into* point 3 as positive, and a flow *away* from point 3 as negative. If the rod is warmer at $x = 4$ than at $x = 3$, that is, if $u(1,4) > u(1,3)$, the rate of flow into point 3 will be (   ) - positive or negative? What is the sign of $u(1,4) - u(1,3)$ here? We express the rate of flow of heat energy in calories per minute.

Now let us work out the *net* rate of flow of heat energy into the point $x = 3$ at the time $t = 1$. This will be the sum of the rates of flow from points 2 to 3 and from points 4 to 5. According to our results so far, this yields

Table 8.1

t	x = 0	1	2	3	4	5
0	0	20	20	20	20	100
1	0					100
2	0					100
3	0	$u(3,1)$	$u(3,2)$	$u(3,3)$	$u(3,4)$	100
4	0					100
5	0					100
6	0					100
7	0					100
8	0					100
9	0					100
10	0					100

net rate of flow of heat energy into $x = 3$ at time $t = 1$

$= k[u(1, \ ) - u(1,3)] + k[u(1,4) - u(1,3)]$
$= k[\qquad]$
$= k[u(1,2) + u(1,4) - 2u(\ , \ )]$

Fill in the missing numbers, factorizing out k for the second line of the expression and simplifying for the third line.

By the definition of specific heat S (look it up in a physics text), the rate of change of temperature at $x = 3$ at the time $t = 1$, that is, the change of temperature per minute is,

$$\frac{u(2,3) - u(1,3)}{1} = \frac{\text{change of temperature, in degrees}}{\text{length of time interval, in minutes}}$$

$$= \frac{1}{S} \cdot (\text{calories per minute of heat flowing into } x = 3)$$

$$= \frac{k}{S} [u(1,2) + u(1,4) - 2u(\ , \ )].$$

More generally, we have the equation

$$\frac{u(t + h,x) - u(t,x)}{h} = \frac{k}{S} [u(t,x - 1) + u(t, x + 1) - 2u(t,x)],$$

which expresses the fact that the rate of change of the temperature, at the time t and the point x, is proportional to the net rate of flow of heat energy into the point x at the time t.

Here k is a positive constant measuring the *heat conductivity* of the material of which the rod is made. We shall assume that the rod is *isotropic* (*k* does not depend on the direction of flow) and *homogeneous* (*k* is the same all along the rod). For most real materials *k* and *S* depend on the temperature $u(t,x)$, but we shall neglect this in our model.

Let us solve the above equation for $u(t + h, x)$:

$$u(t + h, x) = Cu(t, x - 1) + (1 - 2C)u(t,x) + Cu(t, x + 1), \qquad (8.1)$$

where *C* is a certain constant which depends on *h*, *k*, and *S*. What is the value of *C* in terms of these other quantities? If we assign values to *C* (or to *h*, *k*, and *S*) then we can use this equation to *predict* the temperature at the time $t + h$ at the point *x* if we know the state of the rod at the time *t*. We can perform numerical experiments, starting with certain initial and boundary conditions and computing the temperatures at various times and places on the rod.

Exercises

1. Take $k/S = 1/2$, and $h = 1$, and the initial and boundary conditions as:
   $u(0,x) = 20, 0 < x < 5,$
   $u(t,0) = 0, t \geqslant 0,$
   and
   $u(t,5) = 100, t \geqslant 0.$

   Complete table 8.2 of the values of $u(t, x)$ for $0 \leqslant t \leqslant 20, 0 \leqslant x \leqslant 5$.

   Table 8.2

	x =					
t	0	1	2	3	4	5
0	0	20	20	20	20	100
1	0	10	20	20	60	100
2	0	10	15	40	60	100
3	0	7.5	25	37.5	70	100
4	0	12.5	22.5	47.5	68.75	100
5	0	11.25	30	45.625	73.75	100
⋮						

   Carry out your calculations to three decimal places. What seems to be happening to $u(t,1)$ as *t* increases? What seems to be happening to the temperature at each fixed point as time goes on? What does this mean physically?

2. With the same values of k/S and *h* and the same boundary conditions as before, but with the initial conditions
   $u(0,1) = 80, u(0,2) = 60, u(0, 3) = 40, u(0,4) = 20,$

compute a table of the values of $u(t,x)$ for $0 \leqslant t \leqslant 20$. Compare your results with those of the previous exercise.

3. Take the same value of k/S and the same initial and boundary conditions as in exercise 1. Compute $u(1, x)$ for $0 < x < 5$, taking successively $h = 1, h = 0.5$, and $h = 0.1$. What seems to be happening to $u(1,2)$ as h decreases?

4. Suppose that $C = 1/2, u(t,3) \leqslant 60$, and $u(t,4) \leqslant 80$, with the same boundary conditions as before. What is the greatest possible value for $u(t + h, 4)$? Answer the same question if C is any number in the interval $[0, 1]$.

5. We say that the state of the rod is *stationary* if $u(t + h, x) = u(t, x)$ for $0 \leqslant x \leqslant 5$. Suppose that the state is stationary, and that $u(t, 0) = 0, u(t, 5) = 100$. What is the temperature distribution on the rod?

Deviations

Let us return to the problem of solving equation (8.1),

$$u(t + h, x) = Cu(t, x - 1) + (1 - 2C) u(t, x) + Cu(t, x + 1),$$

for

$$x = 1, 2, 3, 4 \text{ and } t \geqslant 0$$

with the boundary conditions

$$u(t,0) = 0, u(t, 5) = 100, \text{ for } t \geqslant 0.$$

In exercises 2 to 5 you found that the only *stationary* state has the temperature distribution

$$u(t,x) = 20x, 0 \leqslant x \leqslant 5.$$

You also found in your numerical experiments that, as t grows larger, the state of the rod seems to approach the stationary state as an *equilibrium* state, no matter what the initial state of the rod may be. We should be able to prove this from our equation (8.1) if our theory is any good.

This suggests that we should describe the state of the rod by means of the *deviation* of the temperature distribution from the stationary distribution:

$$U(t, x) = u(t, x) - 20x.$$

Exercises

6. Prove that $U(t, x)$ satisfies the equation

$$U(t + h, x) = CU(t, x - 1) + (1 - 2C) U(t, x) + CU(t, x + 1)$$

for

$$0 < x < 5, t \geqslant 0,$$

and the boundary conditions

$U(t,0) = U(t,5) = 0$ for $t \geqslant 0$.

7. Predict the behavior of $U(t,x)$ for large t.

Vectors

The state of the rod at the time t would then be described by the numbers

$$U(t,0), \ U(t,1), \ U(t,2), \ U(t,3), \ U(t,4), \ U(t,5),$$

but, by the boundary conditions, the first and last of these numbers are always zero. Therefore it is sufficient to use the *four* numbers $U(t, 1), \ldots,$ $U(t,4)$ to describe the state of the rod. We see, then, that the state of the rod at any time can be described by an ordered *quadruple* of numbers

$$(U_1, U_2, U_3, U_4).$$

(We are using subscripts merely to label the four numbers, and to help us remember which one belongs in which place in the quadruple. The '2' in 'U_2' has nothing to do with the value of U_2. We read this symbol 'U-sub-two.')

It is convenient for us to use a geometric language to talk about these states. This language helps us visualize the algebraic relations between the quadruples of numbers. Just as we can think of an ordered *pair* of numbers, such as $(2, -3)$, as describing a *vector* in two-dimensional space (the plane) (fig. 8.2), so we can think of the ordered quadruple $(1, 2, -3, 4)$ as representing a vector in *four-dimensional space*. Of course, we cannot draw a picture in four dimensions; we can only *imagine* this vector in four-dimensional space.

Fig. 8.2

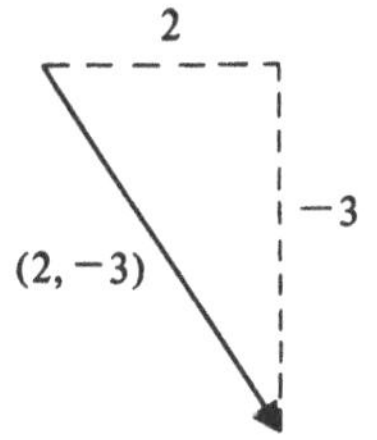

In our problem we shall call the quadruple

$$(U(t, 1)), U(t, 2), U(t, 3), U(t, 4))$$

the *state vector* of the rod at the time t. Equation (8.1) (see exercise 6) tells us how to compute the state vector of the rod at the time $t + h$ if we know the state vector at the time t. Thus equation (8.1) describes the *transition* of the state at any time to the state h minutes later.

Suppose that

$$(U_1, U_2, U_3, U_4)$$

is the state vector at the time t, in other words

$$U(t,1) = U_1, \ldots, U(t,4) = U_4,$$

and, of course, $U(t,0) = U(t,5) = 0$, by our boundary conditions. Then h minutes later we find that

$$U(t + h,1) = CU(t,0) + (1 - 2C)\, U(t,1) + CU(t,2)$$
$$= (1 - 2C)\, U_1 + CU_2,$$
$$U(t + h,2) = (\quad)\, U_1 + (\quad)\, U_2 + (\quad)\, U_3,$$
$$U(t + h,3) = (\quad)\, U_2 + (\quad)\, U_3 + (\quad)\, U_4,$$
$$U(t + h,4) = (\quad)\, U_3 + (\quad)\, U_4.$$

Fill in the missing coefficients.

If (V_1, V_2, V_3, V_4) is the state vector at the time $t + h$, then the equations

$$V_1 = (1 - 2C)U_1 + CU_2,$$
$$V_2 = CU_1 + (1 - 2C)U_2 + CU_3, \text{ etc.,}$$

describe how the state at the time $t + h$ can be computed if you know the state at time t. Write down the equations for V_3 and V_4. We can interpret these equations in a new way. During a time interval of h minutes, the process of heat conduction in the rod *transforms* its state vector (U_1, U_2, U_3, U_4) into the new state vector (V_1, V_2, V_3, V_4), whose components are given by the above equations.

Exercises

8. Express (V_1, V_2, V_3, V_4) in terms of (U_1, U_2, U_3, U_4) in the special case $C = 1/2$.

9. Take $C = 1/2$. Calculate (V_1, V_2, V_3, V_4) for the following choices of (U_1, U_2, U_3, U_4):
    (*a*) $(1, 2, -3, 5)$;
    (*b*) $(2, 4, -6, 10)$;
    (*c*) $(\lambda, 2\lambda, -3\lambda, 5\lambda)$, where λ is any number;
    (*d*) $(7, 6, 5, 4)$;
    (*e*) $(1 + 7, 2 + 6, (-3) + 5, 5 + 4)$;
    (*f*) $(a + 7b, 2a + 6b, -3a + 5b, 5a + 4b)$, where a and b are any numbers.

10. Take $C = 1/2$. Let (W_1, W_2, W_3, W_4) be the state vector at the time $t + 2h$. Give formulas expressing W_1, W_2, W_3, and W_4 in terms of U_1, U_2, U_3, and U_4. If the largest of $|U_1|, |U_2|, |U_3|$, and $|U_4|$ is M, what is the greatest possible value of $|W_1|$, and of $|W_2|$? Give an

example of state vectors (U_1, U_2, U_3, U_4) with $M = 1$, for which your estimate of $|W_2|$ is attained.

11. Suppose $C = 1/2$ and

$$U_1 = 1, U_2 = 2, U_3 = 4\lambda^2 - 1, U_4 = 8\lambda^3 - 4\lambda.$$

Calculate the state vector (V_1, V_2, V_3, V_4) and the vector

$$(V_1 - \lambda U_1, V_2 - \lambda U_2, V_3 - \lambda U_3, V_4 - \lambda U_4).$$

For which values of λ is this last vector the zero vector?

12. If we divided the rod into ten segments of equal length, what would be the dimensionality of the space of the state vectors?

   (a) What is the stationary state for the boundary conditions $u(t,0)$ = 0, $u(t,10) = 100$, for $t \geqslant 0$? Note that we are now taking $L = 10$.

   (b) Give the equations describing the transition of the state vector at any time to the state vector h minutes later. How are these equations simplified if $C = 1/2$?

Vector algebra

We can understand the meaning of our equations better if we intro-duce some elementary vector algebra. It will often be convenient for us to use single letters, such as **U**, as names for vectors. It will help us remember things more easily if we use the symbols

$$U_1, U_2, U_3 \text{ and } U_4$$

to stand for the corresponding components of the vector **U**. In the same way, the vector **V** is the vector with the components

$$V = (V_1, V_2, V_3, V_4).$$

In two dimensions we can add vectors by their components (fig. 8.3):

$$(2, -3) + (5,1) = (2 + 5, (-3) + 1) = (7, -2).$$

Fig. 8.3

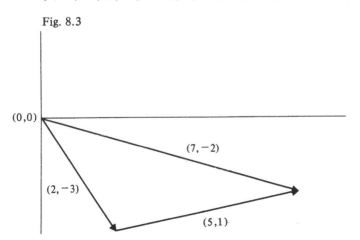

This suggests that we *define* the addition of four-dimensional vectors by the formula

$$U + V = (U_1 + V_1, U_2 + V_2, U_3 + V_3, U_4 + V_4).$$

Thus

$$(1, 2, -3, 5) + (7, 6, 5, 4) = (\quad , \quad , \quad , \quad).$$

Fill in the missing numbers.

We also notice that in two dimensions we can multiply a vector by a number by components (fig. 8.4):

$$3(1,2) = (3 \times 1, 3 \times 2) = (3,6)$$

Fig. 8.4

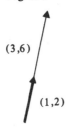

(3,6)

(1,2)

This suggests that we *define* multiplication of the four-dimensional vector U by the number s by the equation

$$sU = (sU_1, sU_2, sU_3, sU_4)$$

Thus

$$2(1, 2, -3, 5) = (\quad , \quad , \quad , \quad).$$

Fill in the missing numbers. We can think of $2U$ as a vector in the same direction as U but twice as long.

The geometric language helps us to interpret the algebraic relations between quadruples of numbers. Our geometric intuition from two and three dimensions suggests what ought to be true in spaces of higher dimension.

Exercises

13. Compute the following vectors:
    (*a*) $(8,8,2,9) + (2,-1,0,1)$.
    (*b*) $(7,6,5,4) + (2,-1,0,1)$.
    (*c*) $(1,2,-3,5) + (9,5,5,5)$.
    (*d*) $(7,6,5,4) + (1,2,-3,5)$.
    (*e*) $4(2,4,-6,10)$.
    (*f*) $8(1,2,-3,5)$.
    (*g*) $(1,2,-3,5) + (-1)(1,2,-3,5)$.

(*h*) 2(7,6,5,4).

(*i*) 2(8,8,2,9).

(*j*) (2,4,−6,10) + (14,12,10,8).

(*k*) $C(U_2,U_1 + U_3,U_2 + U_4,U_3) + (1 - 2C)(U_1,U_2,U_3,U_4)$.

(*l*) 1(1,0,0,0) + 2(0,1,0,0) + (−3)(0,0,1,0) + 5(0,0,0,1).

14. Solve these equations for the unknown vector **U**:

(*a*) (1,2,−3,5) + **U** = (1,2,−3,5).

(*b*) (1,2,−3,5) + **U** = (0,0,0,0).

(*c*) $(U_2,U_1 + U_3,U_2 + U_4,U_3) = \lambda \mathbf{U}$, given that $U_1 = 1$. What are the possible values of λ?

(*d*) $\mathbf{V} = \left(\dfrac{U_2}{2}, \dfrac{U_1 + U_3}{2}, \dfrac{U_2 + U_4}{2}, \dfrac{U_3}{2} \right),$

where $\mathbf{V} = (V_1, V_2, V_3, V_4)$ is any given vector.

Linear transformations

According to exercise 8, when $C = 1/2$ the state vector **V** at the time $t + h$ is obtained from the state vector **U** at the time t by *transforming* **U** as described by the equations

$$\begin{cases} V_1 = & \tfrac{1}{2}U_2 \\ V_2 = \tfrac{1}{2}U_1 & + \tfrac{1}{2}U_3, \\ V_3 = & \tfrac{1}{2}U_2 & + \tfrac{1}{2}U_4, \\ V_4 = & \tfrac{1}{2}U_3. \end{cases}$$

It is convenient to use a single letter, such as T, to stand for such a transformation.

If the transformation T is defined by the above equations, then we can write

$$\mathbf{V} = T\mathbf{U}$$

to show that **V** is the result of performing the transformation T on the vector **U**. For example,

$$T(1,2,-3,5) = (1,-1,7/2,-3/2)$$

(See exercise 9(*a*)). You could express your answer to exercise 9(*d*) in the form

$$T(7,6,5,4) = (\quad , \quad , \quad , \quad).$$

Fill in the blanks.

The transformation T is a special kind of transformation. You have obtained examples of its special properties in exercise 9:

$$T[\lambda(1,2,-3,5)] = \lambda T(1,2,-3,5),$$
$$T[(1,2,-3,5) + (7,6,5,4)] = T(1,2,-3,5) + T(7,6,5,4).$$

More generally, if **X** and **Y** are any four-dimensional vectors and λ is any number then

$$T(\lambda \mathbf{X}) = \lambda T\mathbf{X}, \tag{8.2}$$

and

$$T(\mathbf{X} + \mathbf{Y}) = T\mathbf{X} + T\mathbf{Y}. \tag{8.3}$$

As an illustration, we shall go through the proof of (8.2). If $\mathbf{X} = (X_1, X_2, X_3, X_4)$, then what does $\lambda \mathbf{X}$ mean? Of course we have

$$\lambda \mathbf{X} = (\lambda X_1, \quad , \quad , \quad).$$

Fill in the missing values here and in the following text and exercises. $T(\lambda \mathbf{X})$ is the result of performing the transformation T on the vector $\lambda \mathbf{X}$. The equations defining T tell us how to obtain each component of $T\mathbf{U}$ from the components of **U**. For example, the first component of $T\mathbf{U}$ is one half of the second component of **U**. If $\mathbf{U} = \lambda \mathbf{X}$, then $U_2 = (\quad)$, and therefore the first component of $T(\lambda \mathbf{X})$ is $(1/2)(\quad)$. Proceeding in this way, we see that

$$T(\lambda \mathbf{X}) = (\quad, \tfrac{1}{2}(\lambda X_1 + \lambda X_3), \quad , \quad).$$

In the same way, we can write down the formula for

$$T\mathbf{X} = (\quad, \tfrac{1}{2}(X_1 + X_3), \quad , \quad)$$

and then multiply the result by λ. We then must compare $T(\lambda \mathbf{X})$ with $\lambda T\mathbf{X}$ and see whether they are the same. Write the formula for

$$\lambda T\mathbf{X} = (\quad, \frac{\lambda}{2}(X_1 + X_3), \quad , \quad).$$

Now compare each component of this vector with the corresponding component of $T(\lambda \mathbf{X})$. Are they equal?

If a transformation T satisfies conditions (8.2) and (8.3), then we call T a *linear transformation*. We have thus seen that, for $C = 1/2$, the transformation of the state at time t into the state at time $t + h$ is a linear transformation.

Exercises

15. Prove that T satisfies condition (8.3).

16. Consider the transformation A of two-dimensional vectors defined· by the equations

$$\mathbf{V} = A\mathbf{U}, \text{ where}$$
$$V_1 = U_1 + U_2,$$
$$V_2 = U_1 - U_2.$$

(a) Calculate $A\mathbf{U}$ for the following vectors $\mathbf{U}$:

$\mathbf{U} = (2,3)$,

$\mathbf{U} = (1, \sqrt{2} - 1)$,

$\mathbf{U} = (\frac{1}{2}, \frac{1}{2})$.

(b) Prove that A is a linear transformation.

(c) Find a vector $\mathbf{U}$ such that $A\mathbf{U} = (0,1)$.

(d) Solve the above equations for $\mathbf{U}$ in terms of $\mathbf{V}$:

$U_1 = (\quad)V_1 + (\quad)V_2$,

$U_2 = (\quad)V_1 + (\quad)V_2$.

Do these equations define a linear transformation of $\mathbf{V}$ into $\mathbf{U}$?

17. Let B be the transformation of two-dimensional vectors defined by the equations

$V_1 = U_2$,

$V_2 = -U_1$.

(a) Compute the vectors

$B(1,0), B(1,1), B(1,i)$.

(b) Let A be the transformation of exercise 16. Compute $A[B(1,0)]$ and $B[A(1,0)]$. For example, $A[B(1,0)]$ is the result of applying A to the vector $B(1,0)$ which you have just computed.

(c) If you apply successively the transformations A and B to a vector, does the order of these transformations make any difference?

(d) Find general formulas for $A(B\mathbf{U})$ and $B(A\mathbf{U})$, where $\mathbf{U} = (U_1, U_2)$ is any two-dimensional vector. Call the results $C\mathbf{U}$ and $D\mathbf{U}$ respectively. Are C and D linear transformations? Prove your answer.

(e) Find general formulas for the components of $E\mathbf{U}$, where E is defined by the equation

$E\mathbf{U} = A\mathbf{U} + B\mathbf{U}$.

Is E a linear transformation? Prove your answer.

18. Define the transformation I by the equation $I\mathbf{U} = \mathbf{U}$. I is called the *identity* transformation. Prove that I is a linear transformation.

19. Define the transformation J by the equation

$J\mathbf{U} = 3\mathbf{U}$.

Prove that J is a linear transformation. If A is the transformation of exercise 16, find general formulas for the vectors $A(J\mathbf{U})$ and $J(A\mathbf{U})$. What do you notice?

Algebra of linear transformations

We can operate on linear transformations in various ways to obtain new linear transformations. By using these operations we can create an *algebra* of linear transformations.

For example, if A and B are linear transformations, then the *sum $A + B$* of these transformations is defined by the equation

$$(A + B)U = AU + BU.$$

In exercise 17(c) you found the equations describing $(A + B)U$ for the particular transformations A and B which you studied in exercises 16 and 17. You also proved that $A + B$ is a linear transformation in this special case. It is easy to prove, in general, that if A and B are linear transformations of four-dimensional vectors into four-dimensional vectors then so is $A + B$.

We can also *multiply* linear transformations. The *product AB* of the transformations A and B *taken in that order* is defined by the equation

$$(AB)U = A(BU).$$

In other words, AB is the transformation which consists of first performing B and then applying A to the result. In exercise 17 you found that AB is not always the same as BA. Multiplication of linear transformations *does not*, in general, *have the commutative property*.

We can also multiply a linear transformation by a number:

$$(cA)U = c(AU).$$

For example, $3A$ means the transformation which consists of first performing A and then multiplying the resulting vector by 3.

In creating our algebra of linear transformations, we often have to discuss equality relations between transformations. If A and B are linear transformations of four-dimensional vectors into four-dimensional vectors, we say that

$$A = B$$

if

$$AU = BU$$

for every four-dimensional vector U. Thus one transformation is equal to another if they have the same effect on every vector. It is natural to define powers of linear transformations like this:

$$A^2 = AA, A^3 = AA^2, A^4 = AA^3, \text{ etc.}$$

If T is the transformation of the state vector at the time t into the state vector at the time $t + h$ when $C = 1/2$ (see exercise 8) then we can describe the process of heat conduction by the equation

$$U(t + h) = TU(t), \tag{8.4}$$

where $U(t)$ and $U(t + h)$ are the state vectors at the times t and $t + h$ respectively:

$$U(t) = [U(t,1), U(t,2), U(t,3), U(t,4)],$$

and

$$U(t + h) = [U(t + h, 1), \quad , \quad , \quad].$$

The vector $U(0)$ is the initial state vector.

Compare equation (8.4) with the equation describing a geometric progression:

$$x_{t+1} = rx_t.$$

In a geometric progression, each number is obtained from the previous one by multiplying by a constant *number r*, the common ratio. In our heat conduction problem, each vector is obtained from the previous one by applying a constant *linear transformation.*

We can apply equation (8.4) to calculate the states at the times $0, h, 2h, 3h, \ldots$:

$$U(h) = TU(0),$$

$$U(2h) = TU(h),$$

$$U(3h) = TU(2h), \text{ etc.}$$

From the expression of $U(2h)$ in terms of $U(h)$ and the expression of $U(h)$ in terms of $U(0)$, we obtain an expression of $U(2h)$ in terms of $U(0)$:

$$U(2h) = T[TU(0)] = T^2 U(0).$$

In other words, $U(2h)$ is obtained by applying the transformation T to the initial state $U(0)$, and then applying T again to the result. In the same way, we obtain

$$U(3h) = T(\quad) = T^? U(0)$$

(fill in the missing value), and

$$U(4h) = T^? U(0), \text{ etc.,}$$
$$U(nh) = T^? U(0).$$

What are the missing exponents?

We see that if we wish to know how the state of the rod behaves as time goes on, we must study the behavior of high powers of the transformation T.

Exercises

In exercises 20–24 below we shall be dealing with linear transformations of two-dimensional vectors into two-dimensional vectors defined by the following equations:

$A(U_1,U_2) = (U_1 + U_2, U_1 - U_2)$,

$B(U_1,U_2) = (U_2, - U_1)$,

$D(U_1,U_2) = (U_1, -U_1)$,

$E(U_1,U_2) = (U_1 - U_2, 2U_1 + 4U_2)$.

20. Fill in the blanks to obtain true statements:

(a) $(A + B)(1,0) = A(1,0) + B(1,0) = (\ ,\ ) + (\ ,\ ) = (\ ,\ )$.

(b) $(AB)(1,0) = A[B(1,0)] = A(\ ,\ ) = (\ ,\ )$.

(c) $(5D)(1,0) = 5[D(1,0)] = 5(\ ,\ ) = (\ ,\ )$.

(d) $(B + A)(1,0) = (\ ,\ )$.

(e) $[D(AB)](1,0) = D[(AB)(1,0)] = D(\ ,\ ) = (\ ,\ )$.

(f) $[(DA)B](1,0) = (DA)[B(1,0)] = (DA)(\ ,\ ) = (\ ,\ )$.

(g) $A^2(1,0) = (\ ,\ )$.

(h) $(B - 2I)(1,0) = (\ ,\ )$.

(i) $[DE + (-1)ED](1,0) = (\ ,\ )$.

(j) $A^2(U_1,U_2) = A(U_1 + U_2, U_1 - U_2) = (\ ,\ )$.

(k) $A^3(U_1,U_2) = (\ ,\ )$.

(l) $B^5(U_1,U_2) = (\ ,\ )$.

(m) $D^2(U_1,U_2) = (\ ,\ )$.

21. Solve the following equations for the unknown vector **V**:

(a) $A(0,1) + \mathbf{V} = (1, -1)$.

(b) $A\mathbf{U}^* + V = A\mathbf{U}^*$, where $\mathbf{U}^* = (U_1,U_1)$

(c) $A(0,1) + \mathbf{V} = (0,0)$

(d) $A\mathbf{U} + (-1)\mathbf{V} = (0,0)$, where $\mathbf{U} = (U_1,U_2)$

(e) $[E(A + B)](1,0) = (EA)(1,0) + \mathbf{V}$

(f) $[(AB)D](1,0) = A\mathbf{V}$

(g) $[(A + B)E](0,1) = (AE)(0,1) + B\mathbf{V}$.

22. In each of the following find a linear transformation X such that the equation is true. You may express your answer either in the form

$X\mathbf{U} = (V_1, V_2)$,

where V_1 and V_2 are expressions in terms of U_1 and U_2 such as $U_1 + 3U_2$, or in a form such as

$X = AE + BD + 5I$

or X = some other algebraic combination of the transformations A, B, D, E, and I (see exercise 18).

(a) $A(BD) = XD$.

(b) $A + B = B + X$.

(c) $(A + B) + D = A + X$.

(d) $E(A + B) = EA + X$.

(e) $(A + B)E = AE + X$.

(f) $3(A + B) = 3A + X$.

(g) $(3 + 4)A = 3A + X$.

(h) $A + X = A$.

(i) $A + (-1)X = 0I$.

(j) $AX = I$.

(k) $XA = I$.

(l) $DX = I$.

(m) $(A + B)^2 = A^2 + AB + X + B^2$.

23. In each of the following, find numbers x and y such that the equation is true:

(a) $A^2 + xA + yI = 0I$.

(b) $B^2 + xB + yI = 0I$.

(c) $D^2 + xD + yI = 0I$.

(d) $E^2 + xE + yI = 0I$.

24. In each of the following, find a number x and a *non-zero* vector $\mathbf{U}$ such that the equation is true:

(a) $A\mathbf{U} = x\mathbf{U}$.

(b) $B\mathbf{U} = x\mathbf{U}$.

(c) $D\mathbf{U} = x\mathbf{U}$.

(d) $E\mathbf{U} = x\mathbf{U}$.

Example: If $A\mathbf{U} = x\mathbf{U}$ and $\mathbf{U} = (U_1, U_2)$ then

$$(U_1 + U_2, U_1 - U_2) = x(U_1, U_2) = (xU_1, xU_2),$$

or

$$U_1 + U_2 = xU_1, \quad U_1 - U_2 = xU_2.$$

By the first equation

$$U_2 = (x - 1)U_1,$$

and by the second

$$U_1 - (x - 1)U_1 = x(x - 1)U_1,$$

or

$$(x^2 - 2)U_1 = 0.$$

Therefore either $U_1 = 0$ or $x^2 - 2 = 0$.

If $U_1 = 0$, then $U_2 = (x - 1)U_1 = 0$, so that $\mathbf{U}$ is the zero vector. Therefore we find that

$$x = (\quad) \text{ or } x = (\quad).$$

Fill in the blanks. If $x = \sqrt{2}$, then

$U_2 = (x - 1)U_1 = (\quad)U_1$.

We can choose for U_1 any non-zero number, for instance $U_1 = 1$, and compute U_2 from this equation. We could give as answers either

$x = \sqrt{2}$ and $\mathbf{U} = (1, \sqrt{2} - 1)$ or $x = (\quad)$ and $\mathbf{U} = (1, \quad)$.

25. Prove that if T is the transformation on p. 243 ($C = 1/2$) then the transformation on p. 238 describing the transition of the state vector at the time t to the state at the time $t + h$, for general C, is simply $2CT + (1 - 2C)I$.

26. Give formulas for $T^2\mathbf{U}$, $T^3\mathbf{U}$, and $T^4\mathbf{U}$, where $\mathbf{U} = (U_1, U_2, U_3, U_4)$, an arbitrary four-dimensional vector.

27. Find numbers x, y, z, and t such that

$T^4 + xT^3 + yT^2 + zT + tI = 0I$.

28. Find a number x and a non-zero vector $\mathbf{U}$ such that

$T\mathbf{U} = x\mathbf{U}$,

How many numbers x are there for which this equation has a *nontrivial* (i.e., a non-zero) solution for $\mathbf{U}$?

Inverses

As you may see from exercise 22, the algebra of linear transformations is very much like the ordinary algebra of numbers. We only have to be careful of two things:

(*a*) AB is not, in general, the same as BA;

(*b*) there are transformations, such as D above, which have no inverse with respect to multiplication.

Since we have to be careful about the order in multiplication, some formulas are a bit more complicated than in ordinary algebra; for example, we have

$$(A + B)^2 = (A + B)(A + B) = (A + B)A + (A + B)B$$
$$= A^2 + BA + AB + B^2.$$

Only when $AB = BA$ can we simplify this further:

$$(A + B)^2 = A^2 + AB + AB + B^2 = A^2 + 2AB + B^2.$$

The identity transformation I plays the role of the number 1. The transformations of the form cI, where c is a number, correspond to ordinary numbers. In particular, the transformation $0I$, which transforms all vectors into the zero vector, plays the role of the number 0. Where there is no chance for confusion, we shall often use the symbol '0' to stand for this transformation. Usually you will be able to tell from the context whether we are referring to the number zero or the transformation $0I$.

As you saw in exercise 22(*i*), the inverse of a linear transformation A with respect to addition is $(-1)A$, which we may call $-A$. Subtraction of linear transformations consists of solving such equations as

$A + X = B$

for the unknown transformations X. The solution is

$X = B - A = B + (-1)A.$

We usually denote the inverse of a linear transformation A with respect to multiplication by 'A^{-1}'. For example, you found that if A is the transformation in exercises 20–24, then A^{-1} is the transformation defined by

$A^{-1}(U_1, U_2) = [(U_1 + U_2)/2, (U_1 - U_2)/2]$

(see exercise 22(*e*)-(*j*). Curiously enough, in exercises 20–24 A^{-1} turns out to be equal to $\frac{1}{2}A$).

It makes no difference whether we define A^{-1} by the equation $AX = I$ or the equation $YA = I$. For example, suppose A^{-1} is defined by the first equation. Then we have

$AA^{-1} = I.$

Suppose that Y is a solution of the second equation:

$YA = I.$

Multiply both sides of this equation, on the right, by A^{-1}:

$(YA)A^{-1} = IA^{-1}.$

Now we use the (   ) property of multiplication (which property?) and the fact that I is the identity with respect to multiplication of linear transformations:

$Y(AA^{-1}) = A^{-1}.$

As the factor AA^{-1} is equal to (   ) we conclude that

$Y = YI = A^{-1}.$

So if the first equation has a solution X, which we call A^{-1}, then the only possible solution of the second equation is $Y = A^{-1}$. It is a little more difficult to prove that if the first equation has a solution then so does the second. We shall not discuss this point here.

A linear transformation which has an inverse with respect to multiplication is called *non-singular*. If A is non-singular then we can divide by A, but we must be careful about the order. Thus the solution of the equation

$AX = B$

is

$X = A^{-1}B$

(check it), but the solution of the equation

$$YA = B$$

is

$$Y = (\quad).$$

Fill in the missing value.

In our heat conduction problem, we apply the transformation T to the state vector at the time t to predict the state h minutes in the *future*. We can apply the transformation T^{-1} to find the state h minutes in the *past*.

Exercises

29. Fill in the blanks in the following proof that if A and B are any linear transformations of two-dimensional vectors into two-dimensional vectors then

    $$A + B = B + A.$$

    Proof: let U be any two-dimensional vector. Then, by the definition of $A + B$, we have

    $$(A + B)U = (\quad) + (\quad),$$

    and

    $$(B + A)U = (\quad) + (\quad).$$

    As by the commutative property of addition of vectors

    $$(AU) + (\quad) = (BU) + (\quad),$$

    therefore

    $$(A + \ )U = (B + \ )U$$

    for every two-dimensional vector U. This is, by definition of the equality of transformations, what we mean by

    $$A + B = (\quad) + (\quad).$$

30. Write out proofs of the theorems that if A, B, and C are any linear transformations of two-dimensional vectors into two-dimensional vectors then

    (a) $(A + B) + C = A + (B + C)$;
    (b) $(AB)C = A(BC)$;
    (c) $A(B + C) = AB + AC$;
    (d) $(B + C)A = BA + CA$.

31. Which of the transformations B, C, D and E in exercises 20–24 have inverses with respect to multiplication? If the inverse exists, give equations describing what it does to any two-dimensional vector U.

32. Suppose that the transformation H is defined by the equation

$H(U_1,U_2) = (aU_1 + 2U_2, 3U_1 + dU_2)$.

where a and d are numbers. Prove that H is linear, and that H is non-singular if $ad \neq 6$. If $ad \neq 6$, find a formula for $H^{-1}(U_1,U_2)$.

33. What is the condition on the numbers a, b, c and d that the transformation K defined by

$K(U_1,U_2) = (aU_1 + bU_2, cU_1 + dU_2)$

have an inverse with respect to multiplication? If the inverse exists, find a formula for

$K^{-1}(U_1, U_2)$.

34. For which values of the number C is the transformation

$2CT + (1 - 2C)I$,

of four-dimensional vectors into four-dimensional vectors, non-singular? When it is non-singular, find a formula for

$[2CT + (1 - 2C)I]^{-1} (U_1,U_2,U_3,U_4)$.

35. Prove that if A is a non-singular transformation of two-dimensional vectors into two-dimensional vectors, and if $AU = (0,0)$ then $U = (0,0)$.

Base vectors

As you may have noticed already, it is not easy, in general, to give a simple formula for the nth power of a given linear transformation. It turns out that the problem is simpler if we use the proper coordinate system to describe the vectors and the transformations.

Let us recall briefly the geometric meaning of a coordinate system in the plane. When we wish to label each point or vector by an ordered pair of real numbers, we take any pair of perpendicular lines, and then label each point by its directed distances from these lines (fig. 8.5). We can think of the

Fig. 8.5

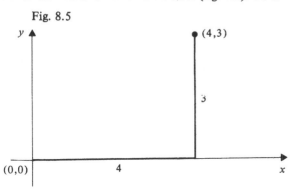

resolution of a vector into components as the process of choosing *base vectors* (1,0) and (0,1), one base vector in the direction of each coordinate axis, and then expressing any other vector, say (4,3), as the sum of vectors in the direction of the base vectors (fig. 8.6):

$$(4,3) = 4(1,0) + 3(0,1)..$$

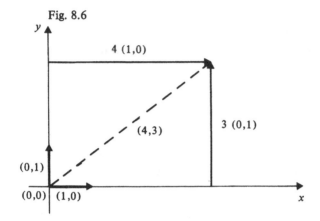

Fig. 8.6

The coordinate system is only a tool for the description of geometric relations. The relations between geometric figures, such as congruence or similarity, are independent of the coordinate system. Indeed we may regard geometry as the study of the properties of geometric figures which are left unchanged, or *invariant*, by changes of the coordinate system.

For example, in the situation shown in figs. 8.5 and 8.6 we might choose two other vectors, say $(1,-1)$ and $(-1,2)$, as base vectors for our coordinate system. We would then describe any other vector, such as (4,3), as a sum of vectors in the directions of these base vectors. In fig. 8.7 we have let $\xi(1,-1)$, where the Greek letter ξ stands for an unknown number, denote the *component* of (4,3) in the direction of the base vector $(1,-1)$. Similarly, we have indicated the component in the direction of the base vector $(-1,2)$ by $\eta(-1,2)$. Make a careful geometric construction on graph paper and estimate (or measure) the values of ξ and η.

We can also find ξ and η algebraically. We want ξ and η to satisfy

$$(4,3) = \xi(1,-1) + \eta(-1,2)$$

$$= (\xi,-\xi) + (-\eta, 2\eta)$$

$$= (\xi - \eta, -\xi + 2\eta).$$

We thus obtain the simultaneous equations

Fig. 8.7

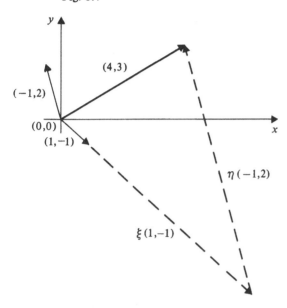

$$\xi - \eta = 4,$$
$$-\xi + 2\eta = 3,$$

by equating corresponding components in the base vectors $(1,0)$ and $(0,1)$. On solving these we find that

$$\xi = (\quad) \text{ and } \eta = (\quad).$$

Fill in the missing values here and subsequently.

More generally, we can pass from our old coordinate system to the new one by solving the equations

$$(x,y) = \xi(1,-1) + \eta(-1,2)$$

for ξ and η in terms of x and y:

$$\xi = (\quad)x + (\quad)y,$$
$$\eta = (\quad)x + (\quad)y.$$

By means of these equations and the corresponding ones for x and y in terms of ξ and η,

$$x = 1\xi + (-1)\eta,$$
$$y = (\quad)\xi + (\quad)\eta,$$

we can easily pass back and forth from one coordinate system to the other.

To illustrate how a certain coordinate system may be especially useful in studying a linear transformation, let us look at the transformation E of exercises 20–24 which is defined by the equation

$E(x,y) = (x - y, 2x + 4y)$.

How is this transformation described in our new coordinate system? We can apply the linearity of E:

$$E[\xi(1,-1) + \eta(-1,2)] = E[\xi(1,-1)] + E[\eta(-1,2)]$$

by property (8.3). Therefore

$$E[\xi(1,-1) + \eta(-1,2)] = \xi E(1,-1) + \eta E(-1,2)$$

by property (8.2). Now we only need to compute

$$E(1,-1) = (\quad,\quad) \text{ and } E(-1,2) = (\quad,\quad)$$

and then express these vectors in our new coordinate system:

$$E(1,-1) = (\quad)(1,-1) + (\quad)(-1,2),$$
$$E(-1,2) = (\quad)(1,-1) + (\quad)(-1,2).$$

We substitute these expressions for $E(1,-1)$ and $E(-1,2)$ in the above equation and obtain

$$E[\xi(1,-1) + \eta(-1,2)] = 2\xi(1,-1) + (\quad)(-1,2).$$

We see that in the new coordinate system, E operates on a vector by multiplying the coefficient of the base vector $(1,-1)$ by $(\quad)$ and the coefficient of the base vector $(-1,2)$ by 3.

Suppose, for the moment, we use the symbol $\langle \xi, \eta \rangle$ to denote the vector $\xi(1,-1) + \eta(-1,2)$, so that

$$(x,y) = \langle 2x + y, x + y \rangle$$

and

$$\langle \xi, \eta \rangle = (\xi - \eta, -\xi + 2\eta)$$

would be another way of describing the relationship between the two coordinate systems. Then in the new coordinate system, we can describe E by the equation

$$E \langle \xi, \eta \rangle = \langle 2\xi, 3\eta \rangle.$$

In which coordinate system is the description of E simpler?

It is now easy to compute

$$E^2 \langle \xi, \eta \rangle = \langle (\quad)\xi, (\quad)\eta \rangle,$$
$$E^3 \langle \xi, \eta \rangle = \langle (\quad), (\quad) \rangle,$$

and in general

$$E^n \langle \xi, \eta \rangle = \langle (\quad), (\quad) \rangle.$$

By using the equations for going from one coordinate system to the other, we can now obtain a formula for E^n in the original coordinate system:

$E^n(x,y) = (\quad , \quad).$

The description in the new coordinate system is still much easier to work with.

Eigenvectors

The explanation of our success in studying E by using the new coordinate system lies in exercise 24. We used as base vectors in our coordinate system those vectors U whose direction is not changed by the transformation E. These vectors are the *eigenvectors* of E.

A non-zero vector U is called an eigenvector of the transformation E if EU is a numerical multiple of U. The multiplier is called the *eigenvalue* which corresponds to that eigenvector. We also say that U is an eigenvector *belonging to* that eigenvalue. For example, $(1,-1)$ is an eigenvector of E belonging to the eigenvalue 2, since $E(1,-1) = 2(1,-1)$. Similarly, $(-1,2)$ is an eigenvector of E belonging to the eigenvalue ( ).

One of the most important mathematical tools in modern physics is the study of linear transformations, their eigenvectors, and their eigenvalues. Indeed in modern atomic physics we think of the state of a physical system as represented by a vector, usually in an infinite-dimensional space. Suppose we want to measure an observable, such as the x-coordinate of the position of an electron. In order to make a measurement of the observable when the system is in a given state, we must *interact* with the system, say by bouncing a light ray off the system and picking up the reflected ray. This interaction transforms the state of the system into a new state. Therefore we think of an observable as represented by a linear transformation acting on the state vector of the system.

In general, when we measure an observable we may obtain any one of a certain set of numbers. When the system is in a given state, there is a certain probability distribution of the possible values of the observable. Only when the state vector is an eigenvector of the transformation corresponding to the observable is the result of the measurement a single definite number. In that case the eigenvalue is the only possible value of the observable. A state whose state vector is an eigenvector of the observable is often called a *pure* state.

Exercises

36. (a) If n is large, which is larger, 2^n or 3^n? What happens to their ratio $2^n/3^n$ as n increases?

   (b) Give an approximate formula for $3^{-n}E^n(x,y)$ for large n. See exercises 20-24.

37. (a) Find the eigenvalues of the transformation A (see exercises 20-24), and find corresponding eigenvectors.

(b) Give a formula expressing the vector (x,y) in a coordinate system whose base vectors are eigenvectors of A.

(c) Give a formula for $A(x,y)$ in terms of this new coordinate system.

38. (a) Calculate $B(1,0)$ and $B(3,4)$, where B is defined as for exercises 20–24, and draw these vectors and the original vectors $(1,0)$ and $(3,4)$ on graph paper. How does the length of each vector compare with the length of its transform by B? What is the angle between any vector and its transform by B? Describe geometrically what B does to any vector.

(b) What are the eigenvalues and eigenvectors of B? (See exercises 20–24.) Represent the eigenvalues of B in the complex plane. Can you draw the eigenvectors of B?

(c) Discuss the questions of parts (a) and (b) of this exercise for the transformation R defined by

$$R(U_1,U_2) = \left(\frac{3}{5} U_1 + \frac{4}{5} U_2, -\frac{4}{5} U_1 + \frac{3}{5} U_2 \right).$$

Conjecture a general principle concerning the eigenvalues of this type of transformation.

39. Find the eigenvalues and corresponding eigenvectors of the transformation S of three-dimensional vectors defined by

$$S(U_1,U_2,U_3) = (2U_1 - U_3, U_2, U_3).$$

How many eigenvalues does S have? Describe geometrically the set of eigenvectors belonging to each eigenvalue.

40. Find the eigenvalues and eigenvectors of the transformation Q defined by

$$Q(U_1,U_2) = (U_1 + U_2, U_1).$$

How many eigenvalues does Q have? Describe geometrically the set of its eigenvectors. Is there a coordinate system in the plane whose base vectors are eigenvectors of Q? Give reasons for your answer.

41. Can the vectors $(1,0)$ and $(2,0)$ be the base vectors for a coordinate system in the plane? Can the vector $(0,1)$ be expressed as a sum of multiples of these two vectors,

$$(0,1) = \xi(1,0) + \eta(2,0)?$$

When can a pair of vectors be used as base vectors for a coordinate system in the plane?

8.2 Cryptography II: a class project

As we have seen, the methods used in the previous discussion of cryptography (section 4.3) are easy to break if one has a large enough sample to analyze statistically. This leads to the search for methods which are harder to break. We still want methods which do not require too much memorizing, or records, or special equipment like coding and decoding machines which might be stolen. It is also desirable to use a method which makes it easy to code or decode with a computer.

Most methods of this sort start by representing the letters of the alphabet by numbers, for example:

A B C D E F G H I J K L M N O P Q R S T U V W X Y Z
0 1 2 3 4 5 6 7 8 9 10 11 12 13 14 15 16 17 18 19 20 21 22 23 24 25

$$(8.5)$$

Next we code pairs of letters instead of single letters. If the message has an odd number of letters, begin by adding any letter, say X, at the end. We may write the second half of the message in reverse under the first half to obtain an arrangement of the numbers in pairs.

Thus if the message is

    CONVOY LEAVES TUESDAY

We add X since there is an odd number of letters and write

    C O N V O Y L E A V
    X Y A D S E U T S E

and translate into numbers:

$$\begin{pmatrix} 2, & 14, & 13, & 21, & 14, & 24, & 11, & 4, & 0, & 21 \\ 23, & 24, & 0, & 3, & 18, & 4, & 20, & 19, & 18, & 4 \end{pmatrix} \qquad (8.6)$$

This gives us a sequence of columns $(2, 23)$, $(14, 24)$, $(13, 0)$, etc.

We now use the array of numbers

$$\begin{pmatrix} 3 & 4 \\ 2 & 3 \end{pmatrix} \qquad (8.7)$$

to transform each column into a new pair according to the following pattern:

$$\left. \begin{aligned} \begin{pmatrix} 3 & 4 \\ 2 & 3 \end{pmatrix} \begin{pmatrix} 2 \\ 23 \end{pmatrix} &= \begin{pmatrix} 3 \times 2 + 4 \times 23 \\ 2 \times 2 + 3 \times 23 \end{pmatrix} \\ &= \begin{pmatrix} 6 + 92 \\ 4 + 69 \end{pmatrix} \\ &= \begin{pmatrix} 98 \\ 73 \end{pmatrix}. \end{aligned} \right\} \quad (8.8)$$

We then divide the resulting numbers by 26, and record the *remainders*:

$$\begin{pmatrix} 98 \\ 73 \end{pmatrix} = \begin{pmatrix} 20 \\ 21 \end{pmatrix}.$$

The rule in (8.8) is *multiply rows by columns.* Note that the numbers in the first row are multiplied by the numbers in the single column, the products are added, and the sum is in the first row of the result. In the same way, the result of multiplying the second row by the column goes in the second row of the transformed column. We give here the transformation of the second column:

$$\begin{pmatrix} 3 & 4 \\ 2 & 3 \end{pmatrix}\begin{pmatrix} 14 \\ 24 \end{pmatrix} = \begin{pmatrix} 138 \\ 100 \end{pmatrix} = \begin{pmatrix} 8 \\ 22 \end{pmatrix}.$$

Exercises

42. (a) Transform the rest of the array in (8.6).
    (b) Translate back into letters, and rearrange as before. This is the coded message.
43. Which letters are repeated in the plain text? Are they repeated in the coded text? Which letters are repeated in the coded text? What are the corresponding plain-text letters? Compare the two codings of the repeated digraph ES.
44. Make up a short message of your own. Encode it by the above method. Exchange coded messages with a classmate. Can you figure out how to decode your friend's message?
45. Use the array

$$\begin{pmatrix} 2 & 5 \\ 1 & 2 \end{pmatrix}$$

to encode the above message or any other message by the above method.

An array of numbers arranged in a rectangle is called a matrix. The array (8.6) is called a 2 × 10 matrix, since it has 2 rows and 10 columns. We used the 2 × 2 matrix in (8.7) to transform (8.6) into the new 2 × 10 matrix you obtained in exercise 42. We can express this relation as follows:

$$\begin{pmatrix} 3 & 4 \\ 2 & 3 \end{pmatrix}\begin{pmatrix} 2 & 14 & 13 & 21 & 14 & 24 & 11 & 4 & 0 & 21 \\ 23 & 24 & 0 & 3 & 18 & 4 & 20 & 19 & 18 & 4 \end{pmatrix}$$
$$= \begin{pmatrix} 98 & 138 & 39 & 75 & 114 & \dots \\ 73 & 100 & 26 & 51 & 82 & \dots \end{pmatrix}.$$

Write in the missing numbers. In the same way, we can write the transformation of the individual columns, for example, (8.8) above or

$$\begin{pmatrix} 3 & 4 \\ 2 & 3 \end{pmatrix}\begin{pmatrix} 24 \\ 4 \end{pmatrix} = \begin{pmatrix} 88 \\ 60 \end{pmatrix}.$$

A column of two numbers may be thought of as a 1 × 2 matrix.

We see that in order to carry out the multiplication, the first matrix must have as many columns as the second matrix has rows. Thus we can multiply a 3 × 2 matrix by a 2 × 4 matrix and obtain a 3 × 4 matrix:

$$\begin{pmatrix} 1 & 2 \\ 3 & 4 \\ 5 & 6 \end{pmatrix} \begin{pmatrix} 1 & 2 & 3 & 4 \\ 5 & 6 & 7 & 8 \end{pmatrix} = \begin{pmatrix} 11 & 14 & 17 & 20 \\ 23 & 30 & 37 & 44 \\ 35 & 46 & 57 & 68 \end{pmatrix}.$$

We shall work here mostly with multiplying by 2 × 2 matrices.

Exercises

46. Multiply out the following matrices:

(a) $\begin{pmatrix} 1 & 0 \\ 0 & 1 \end{pmatrix} \begin{pmatrix} 3 & 4 \\ 5 & 6 \end{pmatrix}$

(b) $\begin{pmatrix} 3 & 4 \\ 5 & 6 \end{pmatrix} \begin{pmatrix} 1 & 0 \\ 0 & 1 \end{pmatrix}$

(c) $\begin{pmatrix} 1 & 1 \\ 0 & 1 \end{pmatrix} \begin{pmatrix} 1 & 0 \\ 1 & 1 \end{pmatrix}$

(d) $\begin{pmatrix} 1 & 0 \\ 1 & 1 \end{pmatrix} \begin{pmatrix} 1 & 1 \\ 0 & 1 \end{pmatrix}$

(e) $\begin{pmatrix} 1 & 0 \\ 1 & 1 \end{pmatrix} \begin{pmatrix} 2 & 1 \\ 1 & 1 \end{pmatrix}$

(f) $\begin{pmatrix} 1 & 1 \\ 1 & 2 \end{pmatrix} \begin{pmatrix} 1 & 0 \\ 1 & 1 \end{pmatrix}$

(g) $\begin{pmatrix} 1 & 0 \\ 1 & 1 \end{pmatrix} \begin{pmatrix} 3 & 2 \\ 1 & 2 \end{pmatrix}$

(h) $\begin{pmatrix} 2 & 2 \\ 1 & 3 \end{pmatrix} \begin{pmatrix} 1 & 0 \\ 1 & 1 \end{pmatrix}$

47. Can you generalize the pattern in exercise 46(a) and (b)? Guess the general law and test your guess.

48. Compare exercise 46(c) and (d). Does the order of factors make a difference in matrix multiplication? Does the commutative law

$AB = BA$

hold for matrix multiplication?

49. In the product of the three factors

$$\begin{pmatrix} 1 & 0 \\ 1 & 1 \end{pmatrix} \begin{pmatrix} 1 & 1 \\ 0 & 1 \end{pmatrix} \begin{pmatrix} 1 & 0 \\ 1 & 1 \end{pmatrix}$$

does it make any difference how the factors are grouped? Test the associative law

$(AB)C = A(BC)$

for other choices of the matrices A, B, and C. Use 2×2 matrices.

50. What is the relation of the pair of matrices

$$\begin{pmatrix} 1 & 1 \\ 0 & 1 \end{pmatrix} \text{ and } \begin{pmatrix} 2 & 1 \\ 1 & 1 \end{pmatrix}$$

to the matrix

$$\begin{pmatrix} 3 & 2 \\ 1 & 2 \end{pmatrix}?$$

What is the relation between the answers in exercise 46(*d*), (*e*), and (*g*)? Guess at the general law and test your guess with other choices of three matrices. Compare with the distributive law

$$A(B + C) = AB + AC$$

for multiplication and addition of numbers.

51. Compare your results in exercise 46(*c*), (*f*), and (*h*). Guess at the general law and test your guess. Compare with the distributive law

$$(B + C)A = BA + CA.$$

52. If

$$Y = \begin{pmatrix} 3 & 4 \\ 2 & 3 \end{pmatrix} X,$$

and

$$Z = \begin{pmatrix} 3 & -4 \\ -2 & 3 \end{pmatrix} Y,$$

find a matrix C such that

$$Z = CX.$$

Try any 2×1 matrix and any 2×2 matrix for X.

53. Try exercise 44 again using exercise 52.

54. How can you decode messages which were encoded with the matrix of exercise 45?

55. How can you decode messages which are encoded with the matrix

$$\begin{pmatrix} 1 & 1 \\ 2 & 3 \end{pmatrix}?$$

As you see, the secret of decoding the code discussed in the text is to find a matrix B such that

$$\begin{pmatrix} 3 & 4 \\ 2 & 3 \end{pmatrix} B = \begin{pmatrix} 1 & 0 \\ 0 & 1 \end{pmatrix}.$$

If B is the unknown matrix

$$B = \begin{pmatrix} x & y \\ u & v \end{pmatrix}$$

then you obtain the equations

$$3x + 4u = 1, 3y + 4v = 0,$$
$$2x + 3u = 0, 2y + 3v = 1.$$

You can solve these equations like this:

$$u = -\frac{2x}{3}, \qquad\qquad v = -\frac{3y}{4},$$

$$3x - \frac{8x}{3} = 1, \qquad\qquad 2y - \frac{9y}{4} = 1,$$

$$9x - 8x = 3, \qquad\qquad 8y - 9y = 4,$$

$$x = 3, \qquad\qquad y = -4,$$

$$u = -2, \qquad\qquad v = 3.$$

This explains the result of exercise 52.

Exercises

56. Try exercises 54 and 55 again.
57. How can you decode messages which were encoded with the matrix

$$\begin{pmatrix} 2 & 5 \\ 1 & 3 \end{pmatrix}?$$

58. Try decoding the message of the text using the matrix

$$\begin{pmatrix} 3 & 22 \\ 24 & 3 \end{pmatrix}.$$

Does it work? If so, can you explain why?

In the above calculations we discarded multiples of 26, so that we had

$$\begin{pmatrix} 138 \\ 100 \end{pmatrix} = \begin{pmatrix} 8 \\ 22 \end{pmatrix}$$

because

$$138 = 8 + (5 \times 26),$$
$$100 = 22 + (3 \times 26).$$

This is called calculating *modulo* 26.

Let us explain calculating modulo 3. We can arrange the whole numbers in the following way:

```
0   1   2
3   4   5
6   7   8
9  10  11
12  13  14  etc.
```

The name of any column is the number at the top.

Add any number in the 1-column to any number in the 2-column. What is the column of the sum? Does it make any difference which numbers you choose in those columns? Try the same experiment with any pair of columns. Does something similar happen? Is

(1 + multiple of 3) + (2 + multiple of 3)

always equal to

(0 + multiple of 3)?

Explain the results of your other experiments in the same way.

Try the same experiments with multiplication instead of addition. For example, what are the columns of the products

4 × 8, 7 × 5, 13 × 2, 10 × 14?

Incidentally, what is a fast way of finding the column of a number like 140?

Exercises

59. Make tables for addition and multiplication modulo 3:

```
+ | 0  1  2        × | 0  1  2
--+--------        --+--------
0 |                0 |
1 |    0           1 |       2
2 |                2 |
```

We have recorded

1 + 2 = 0, 1 × 2 = 2 modulo 3,

which expresses the results of the above experiments.

60. How can you use the above addition table to subtract modulo 3? For example,

1 − 2 = ? modulo 3.

61. Make tables for addition and multiplication modulo 26.

62. Why is dividing by 3 modulo 26 the same as multiplying by 9 modulo 26? Can you divide by 2 modulo 26? What about 5? Make a list of the numbers you can divide by, modulo 26.

63. Make a list of the numbers you can divide by, modulo 3.

64. If you used the matrix

$$\begin{pmatrix} 3 & 4 \\ 4 & 6 \end{pmatrix}$$

to encode messages, could they be decoded? Try the message
CONVOY DEPARTS FRIDAY.

What happens when you transform the columns

$$\begin{pmatrix} 14 \\ 11 \end{pmatrix} \text{ and } \begin{pmatrix} 14 \\ 24 \end{pmatrix}$$

corresponding to the pairs of letters OI and OY? Encode the message
CONVOY DEPARGF SEYQNI

and compare the results.

65. Can you decode messages encoded with the matrix

$$\begin{pmatrix} 3 & 4 \\ 6 & 9 \end{pmatrix}?$$

8.3 Linear algebra in calculus

Differentiation, integration, multiplication

In recent years, authoritative groups such as the Committee on the Undergraduate Curriculum in Mathematics of the Mathematical Association of America have recommended that the first two years of the college program should include a substantial amount of linear algebra. This can be done most efficiently by connecting it with calculus. We give here some illustrations of how this can be done.

The set $\mathcal{P}$ of polynomials in x (with real coefficients) is a vector space over the field $\mathbb{R}$ of real numbers. It has the following properties:

$$\left. \begin{aligned} &\text{If } P \in \mathcal{P}, Q \in \mathcal{P}, R \in \mathcal{P}, \text{ then } P + Q \text{ and } cP \text{ are in } \mathcal{P} \text{ for all } c \in \mathbb{R}, \\ &P + Q = Q + P, (P + Q) + R = P + (Q + R) \\ &a(bP) = (ab)P, \; 0P = 0, \; 1P = P, \; c(P + Q) = cP + cQ. \end{aligned} \right\} \quad (8.9)$$

The operation $\hat{D}$ of differentiating with respect to x is a linear operator on $\mathcal{P}$:

$$\left. \begin{aligned} &\text{If } P \in \mathcal{P}, \text{ then } \hat{D}P \in \mathcal{P}, \\ &\hat{D}(P + Q) = \hat{D}P + \hat{D}Q, \\ &\hat{D}(cP) = c\hat{D}P. \end{aligned} \right\} \quad (8.10)$$

The operation $\hat{J}$ defined by

$$(\hat{J}P)(x) = \int_0^x P(t)\mathrm{d}t$$

is also a linear operator on $\mathscr{P}$. It is a *right inverse* of $\hat{D}$:

$$\hat{D}\hat{J}P = P \text{ for all } P \in \mathscr{P} \tag{8.11}$$

This equation can be expressed as a relation between linear operators:

$$\hat{D}\hat{J} = \hat{1}, \tag{8.11a}$$

where $\hat{1}$ stands for the operation of multiplying by 1, and is the *identity* operator. In the following text we will not distinguish between $\hat{1}$ and 1, as it will be clear from the context which is intended. Is $\hat{J}$ a *left inverse* of $\hat{D}$, that is, does

$$\hat{J}\hat{D} = 1$$

or

$$\hat{J}\hat{D}P = P ?$$

Is this last statement true if $P(x) = x + 5$?

Exercises

66. (a) Does $\hat{D}$ have any other right inverses? Find a right inverse $\hat{J}_1$ of $\hat{D}$ such that

$$\hat{J}_1 x = (x^2 - 1)/2.$$

(b) If $\hat{A}$ is a right inverse of $\hat{D}$, what can $\hat{A}x$ be?

67. Can $\hat{D}$ have *any* left inverse? If $\hat{L}$ were a left inverse, and $\hat{A}$ any right inverse, calculate

$$\hat{L}\hat{D}\hat{A} = (\hat{L}\hat{D})\hat{A} = \hat{L}(\hat{D}\hat{A})$$

in two ways.

There is a natural algebra of linear operators. The basic operations on linear operators $\hat{L}$ and $\hat{M}$ are defined by the following equations, where P is any element of $\mathscr{P}$ and c is in $\mathbb{R}$:

$$\left.\begin{array}{l} (\hat{L} + \hat{M})P = (\hat{L}P) + (\hat{M}P), \\ (\hat{L}\hat{M})P = \hat{L}(\hat{M}P), \\ (c\hat{L})P = c(\hat{L}P). \end{array}\right\} \tag{8.12}$$

We have already used the product of linear operators in discussing $\hat{D}\hat{J}$:

If $P(x) = 3$ then $\hat{J}\hat{D}P = 0$,

but if $P(x) = x$ then $\hat{J}\hat{D}P = P$.

More generally, for any P in $\mathscr{P}$, we have

$$(\hat{J}\hat{D}P)(x) = (\hat{J}P')(x) = \int_0^x P'(t)dt = P(x) - P(0).$$

so that

$$\hat{J}\hat{D} = 1 - \hat{\delta}_0,$$ (8.13)

where $\hat{\delta}_0$ is the operator of *evaluating* at zero,

$$\hat{\delta}_0 P = P(0).$$

Equations (8.11a) and (8.13) show that $\hat{J}$ and $\hat{D}$ do not commute. This multiplication of linear operators is not commutative in general. We can describe the non-commutativity of $\hat{J}$ and $\hat{D}$ more precisely by the equation

$$\hat{D}\hat{J} - \hat{J}\hat{D} = \hat{\delta}_0.$$ (8.14)

The combination

$$[\hat{L},\hat{M}] = \hat{L}\hat{M} - \hat{M}\hat{L}$$ (8.15)

of two linear operators $\hat{L}$ and $\hat{M}$ is called their *commutator*; it measures the extent to which they do not commute. They commute if and only if their commutator is zero. The operator $\hat{X}$ of multiplying by x is defined by

$$(\hat{X}P)(x) = xP(x).$$ (8.16)

The equation

$$[\hat{D},\hat{X}] = 1$$ (8.17)

follows from the rule for differentiating a product.

Equation (8.17) is the mathematical formulation of the *Heisenberg uncertainty relation*, which is fundamental in modern physics. Its physical interpretation is that the position and momentum of an electron cannot be measured precisely and simultaneously.

Exercises

68. Verify equation (8.17).
69. Do $\hat{X}^2$ ($\hat{M}^2$ means $\hat{M}\hat{M}$) and $\hat{X}$ commute? What is

$$[\hat{D}, \hat{X}^2] ?$$

70. What is $\hat{\delta}_0^2$? What is $\hat{\delta}_0\hat{X}$?
71. Compute $\hat{D}(\hat{X}\hat{J} - \hat{J}\hat{X})$ and $\hat{D}\hat{J}^2$. What is the relation between $[\hat{X},\hat{J}]$ and $\hat{J}^2$?
72. Use the idea of exercise 71 to find a simple formula for

$$\hat{X}^2\hat{J} - 2\hat{X}\hat{J}\hat{X} + \hat{J}\hat{X}^2.$$

Because multiplication of operators is not, in general, commutative, one must be careful about the order of multiplication. For example, we have

$$(\hat{D} - \hat{X})(\hat{D} + \hat{X}) = \hat{D}(\hat{D} + \hat{X}) - \hat{X}(\hat{D} + \hat{X})$$

$$= \hat{D}^2 + \hat{D}\hat{X} - \hat{X}\hat{D} - \hat{X}^2$$

$$= \hat{D}^2 + 1 - \hat{X}^2,$$

by (8.17). Contrast this with the identity

$$(a - b)(a + b) = a^2 - b^2$$

in ordinary algebra.

Everything we have done so far works equally well for the set $C^\infty(\mathbb{R})$ of functions f on $\mathbb{R}$ to $\mathbb{R}$ such that $f^{(n)}(x)$ (the nth derivative of f at x) exists for all $x \in \mathbb{R}$ and all n. This is also a vector space, and $\hat{D}$, $\hat{J}$, and $\hat{X}$ are also linear operators on it. We could also work with the vector space $C^\infty[0,1]$ of all functions f on the interval $[0,1]$ to $\mathbb{R}$ such that $f^{(n)}(x)$ exists for all n and all x in $[0,1]$.

The space $C^1(\mathbb{R})$ of functions f on $\mathbb{R}$ to $\mathbb{R}$, such that $f'(x)$ exists and is continuous for all x, is also a vector space over $\mathbb{R}$. The operators $\hat{J}$ and $\hat{X}$ are also linear on this space to itself. The linear operator $\hat{D}$ is a transformation from $C^1(\mathbb{R})$ to $C(\mathbb{R})$, the space of continuous real-valued functions on $\mathbb{R}$.

The *kernel* of $\hat{D}$ in $C^1(\mathbb{R})$ is the set of f which are transformed into zero:

$$\hat{D}f = 0.$$

We see that the kernel of $\hat{D}$ is the set of all constant functions. These may be identified with $\mathbb{R}$ itself.

Given g in $C(\mathbb{R})$, to find all f such that

$$\hat{D}f = g, \tag{8.18}$$

we note that if

$$\hat{D}f_1 = \hat{D}f_2 = g$$

then

$$\hat{D}(f_1 - f_2) = g - g = 0.$$

Hence $f_1 - f_2$ is in the kernel of $\hat{D}$. If f_0 is one solution of (8.18) then we can obtain all solutions of (8.18) in the form

$$f = f_0 + k,$$

where k is an arbitrary element of the kernel of $\hat{D}$.

The finite difference operators are also interesting. The simplest one is Δ, defined by

$$(\Delta f)(x) = f(x + 1) - f(x). \tag{8.19}$$

For example,

$$\Delta x^2 = (x + 1)^2 - x^2 = 2x + 1.$$

This is a linear operator on all the spaces mentioned above except $C^\infty[0,1]$. The function $\sin(2\pi x)$ is in the kernel of Δ in $C(\mathbb{R})$. The algebra generated by the operators $\hat{X}$ and $\hat{D}$ is a simple example of a non-commutative algebra which can be treated at the high school level.

Exercises

73. Compute $(\hat{D} + \hat{X})(\hat{D} - \hat{X})$ and $[\hat{D} + \hat{X}, \hat{D} - \hat{X}]$.
74. Compute $[\Delta, \hat{X}]$ and $[\Delta, \hat{X}^2]$.
75. Compute $[\hat{D}^2, \hat{X}]$ and $[\Delta^2, \hat{X}]$.
76. What is the kernel of $\hat{D}^2$ in $C^1(\mathbb{R})$?
77. Give at least two functions in the kernel of Δ.
78. What is the kernel of $\hat{\delta}_0$ in $C(\mathbb{R})$?
79. (a) What is the kernel of $\hat{J}$ in $C(\mathbb{R})$?
    (b) Given g in $C(\mathbb{R})$, how many solutions does the equation $\hat{J}f = g$ have?
80. What is the kernel of $\hat{X}$ in $C(\mathbb{R})$? Does $\hat{X}$ have a right inverse in $C(\mathbb{R})$? Does it have a left inverse? What about in $\mathscr{P}$?
81. Let $\mathscr{P}_0$ be the kernel of $\hat{\delta}_0$ in $\mathscr{P}$. Find an operator $\hat{A}$, which transforms $\mathscr{P}_0$ into $\mathscr{P}$, such that $\hat{A}\hat{X} = 1$ on $\mathscr{P}$.
82. Compute $[\hat{D}, \hat{X}^3 - 2\hat{X} + 5]$. If $f \in C'(\mathbb{R})$, and $\hat{F}$ denotes the operation of multiplying by $f(x)$,

$$(\hat{F}g)(x) = f(x)g(x),$$

give a simple formula for $[\hat{D}, \hat{F}]$.

Linear differential equations

The differential equation

$$\frac{d^2y}{dx^2} - 3\frac{dy}{dx} + 2y = x^2 + 6 \tag{8.20}$$

can be written in the form

$$(\hat{D}^2 - 3\hat{D} + 2)y = x^2 + 6.$$

We can think of this as the problem of finding all functions in $C^2(\mathbb{R})$ which are transformed by the linear operator $\hat{D}^2 - 3\hat{D} + 2$ into the function $x^2 + 6$. The operator $\hat{D}^2 - 3\hat{D} + 2$ is called a *linear differential operator*. It is said to be of *second* order because it involves the second but no higher derivative.

As in the discussion of (8.18), we note that if y_0 and y are solutions of (8.20) then

$$(\hat{D}^2 - 3\hat{D} + 2)(y - y_0) = 0,$$

so that $y - y_0$ is in the kernel of $\hat{D}^2 - 3\hat{D} + 2$. This kernel is the set of functions k which satisfies the equation

$$(\hat{D}^2 - 3\hat{D} + 2)k = 0 \tag{8.21}$$

One way to solve (8.21) is to factorize the operator:

$$\hat{D}^2 - 3\hat{D} + 2 = (\hat{D} - 2)(\hat{D} - 1)$$
$$= (\hat{D} - 1)(\hat{D} - 2).$$

Hence if

$$(\hat{D} - 1)k = 0 \text{ or } (\hat{D} - 2)k = 0$$

then k satisfies (8.21). The equation

$$(\hat{D} - 1)k = 0 \text{ or } k' = k$$

we recognize as the equation for the exponential function. Its solutions are

$$k(x) = Ae^x,$$

where A is an arbitrary constant. Similarly the solutions of

$$(\hat{D} - 2)k = 0$$

are of the form

$$k(x) = Be^{2x},$$

where B is an arbitrary constant. Hence all functions of the form

$$k(x) = Ae^x + Be^{2x}, A \in \mathbb{R}, B \in \mathbb{R}, \tag{8.22}$$

are in the kernel of $\hat{D}^2 - 3\hat{D} + 2$. Are there any other functions in this kernel?
    To answer this, let us solve (8.21) in another way. If we set $(\hat{D} - 2)k = u$, then the equation can be written in the form of the system

$$\left.\begin{array}{l} (\hat{D} - 2)k = u, \\ (\hat{D} - 1)u = 0. \end{array}\right\} \tag{8.23}$$

The solutions of the second equation here are, of course

$$u(x) = Ae^x, A \in \mathbb{R}.$$

Hence our problem is reduced to that of solving the *first order* differential equation

$$(\hat{D} - 2)k = Ae^x. \tag{8.24}$$

To solve this, we note that again it is sufficient to find *one* solution, so we simply add an arbitrary function in the kernel of $\hat{D} - 2$.
    To find one solution, we note that

$$(\hat{D} - 2)e^x = \hat{D}e^x - 2e^x = -e^x.$$

Since $\hat{D} - 2$ is a linear operator, the function

$$k(x) = -Ae^x$$

is a solution of (8.24). Hence the general solution of (8.24) is

$$k(x) = (-A)e^x + Be^{2x}, B \in \mathbb{R}.$$

If A is also an arbitrary constant, this is the general solution of (8.21). (Note that if A is an arbitrary constant then so is $-A$.)

We can express our result (8.22) in another form. Let $k_1(x) = e^x$, $k_2(x) = e^{2x}$. These are two elements in the kernel of $\hat{D}^2 - 3\hat{D} + 2$. Every element in this kernel is, by (8.22), a *linear combination*

$$k = Ak_1 + Bk_2, A \in \mathbb{R}, B \in \mathbb{R},$$

of these two. If $k = 0$, then $A = B = 0$, which means that k_1 and k_2 are *linearly independent*. These facts can be summarized in the statement

The kernel of $\hat{D}^2 - 3\hat{D} + 2$ is a *two-dimensional subspace* of $C^1(\mathbb{R})$, and k_1 and k_2 constitute a basis for this subspace.

To solve the equation

$$y'' + (1 - x^2)y = x + 2$$

or

$$[\hat{D}^2 + (1 - \hat{X}^2)]\,y = x + 2$$

is somewhat more difficult because the factors of $\hat{D}^2 + 1 - \hat{X}^2$ *do not commute* (exercise 70). We can still approach it by transforming it into a system of two first order equations like (8.23). Again, the kernel of the operator is two-dimensional, but one of the functions in the basis is not an elementary function.

Exercises

83. If y is a polynomial of degree n, is

$$(\hat{D}^2 - 3\hat{D} + 2)y$$

a polynomial? If so, of what degree? Can you find a polynomial solution of (8.20)? If so, of what degree? What is the general solution of (8.20)?

84. (*a*) If $m \in \mathbb{R}$, compute $(\hat{D}^2 - 3\hat{D} + 2)e^{mx}$.

(*b*) Find the general solution of

$$(\hat{D}^2 - 3\hat{D} + 2)y = 4e^{3x}.$$

85. (*a*) If a, b, and m are in $\mathbb{R}$, compute

$$(\hat{D}^2 - 3\hat{D} + 2)(a\cos(mx) + b\sin(mx)).$$

(*b*) Find the general solution of

$$(\hat{D}^2 - 3\hat{D} + 2)y = \cos(3x) + 5\sin(3x).$$

86. Find the general solution of

$$(\hat{D}^2 + 9)y = 7e^x.$$

87. (*a*) Find the general solution of

$$(\hat{D}^3 - 6\hat{D}^2 + 11\hat{D} - 6)y = 0.$$

(*b*) Find the general solution of

$$(\hat{D}^3 - 6\hat{D}^2 + 11\hat{D} - 6)y = 15e^x.$$

88. Suppose we are looking for the solution of (8.20) such that $y(0) = 1$, $y'(0) = -2$. Let $u = \hat{D}^2 y$.

   (*a*) Show that $\hat{J}u - \hat{D}y$ is a constant. Which constant?

   (*b*) Show that $\hat{J}^2 u - y$ is a first degree polynomial. Which polynomial?

   (*c*) Show that (8.20) with the above initial conditions is equivalent to an *integral* equation:

   $$u = 3\hat{J}u + 2\hat{J}^2 u = f(x),$$

   where f is a known function.

Remarks on teaching calculus

In this section we have not attempted to give any applications which lead directly to the algebraic aspects of calculus. We suggest that it is most natural to introduce algebraic concepts and terminology in the teaching of calculus gradually and in connection with mathematical problems.

Thus after studying the concept of differentiation it is natural to discuss the properties listed under (8.9) and (8.10) above. Even before the general problem of integration is discussed, it is natural to look for the polynomial solutions Q of the equation

$$\hat{D}Q = P.$$

We can obtain the general solution from the solution of the special case where $P(x) = x^n$ is a power of x. It is then natural to ask whether there is a linear operator $\hat{L}$ on P such that $Q = \hat{L}P$ is a solution of this equation, that is,

$$\hat{D}\hat{L}P = P.$$

We are thus led to the operator $\hat{J}$, and its properties as embodied in (8.11), (8.11*a*), and (8.13).

After the algebra of linear operators is introduced as indicated in (8.12), commutators of specific linear operators, and then the general concept as in (8.15), arise naturally out of the algebra.

The algebra generated by the operators $\hat{X}$ and $\hat{D}$ is a simple natural example of a non-commutative algebra. As we have indicated, the algebra generated by Δ and $\hat{X}$ can be introduced at the high school level. In many mathematics courses from the ninth grade upwards, there is given at the beginning a review of the properties of the real number system. This is very boring for the students who remember them from previous years, yet it must be included since there are always students who forget them or never learned them. A more motivated way to accomplish this review is to begin with a new algebra, say complex numbers, 2 X 2 matrices, or one of the algebras occurring in this section. Then one can investigate which of the properties of the real number system hold for the new algebra. As a by-product one obtains the desired review.

The next place where new algebraic ideas come up naturally is in the study of linear differential equations.

In most of the applications discussed in previous sections, the passage from the real world to the mathematical model is simple enough and intuitive enough to be presented in a mathematics class. Quantum physics, however, deals with phenomena that lead to a conception of the microscopic world quite different from our experience with the macroscopic world. Since 1956, it has become more and more customary to introduce these conceptions at least by the second college year to science and engineering students, and attempts are being made to present them even at the high school level. If the students are learning or have learned some quantum physics, then one can discuss simple cases of the mathematical models involved.

Suppose we consider a particle, say a photon, moving along a line represented by the x-axis. Then the *state of the particle* is represented in quantum mechanics by a function $\psi(x)$. The values of $\psi(x)$ are, in general, complex numbers. We do not imagine direct relation between the value of ψ at a particular point, say $\psi(2)$, and anything observable. But we do associate certain integrals with experimental results.

For example, imagine that we perform an experiment to find out whether the particle, when it is in the state ψ, is in the interval $(0,3)$. The probability of finding a 'yes' answer is given by the integral

$$P(0 < x < 3) = \int_{0}^{3} |\psi(x)|^2 \, dx.$$

This illustrates a typical difference between the quantum and the classical points of view. In classical physics we would define the state of the particle in terms of a pair of numbers (x,p), the position and momentum. We imagine that these have definite values and can, in principle, be measured as accurately as we please. In quantum mechanics, however, the state is represented by a function ψ. When we measure the position X for a particle in the state ψ, we do not, in general, obtain a single value. If we perform the experiment repeatedly, we obtain values with a certain distribution. If the number of observations is large then the relative frequency with which X occurs in the interval $(0,3)$ will be close to the above integral.

The *expected value* of X (see p. 129) is given by the integral

$$E(X) = \int_{-\infty}^{\infty} x|\psi(x)|^2 dx \qquad (8.25)$$

If we repeatedly measure X when the particle is the state ψ, the average of the values of X will be close to this integral.

In general, an observable A is represented by a linear operator $\hat{A}$ which transforms states into states. The expected value of A when the system is in the state ψ is given by the integral

$$E(A) = \int_{-\infty}^{\infty} \overline{\psi(x)} \, (\hat{A}\psi)(x)dx,$$

where $\overline{\psi(x)}$ is the complex conjugate of $\psi(x)$. Thus the position-observable X is represented by the operator $\hat{X}$ of multiplying by x, as in equation (8.16), and its expectation is given by (8.25). In quantum mechanics we consider only observables whose expectations are real numbers for all possible states. We usually consider only states ψ which are small when x is large. The momentum P is represented by the operator

$$\hat{P} = -i\hat{D},$$

so that its expectation is given by the integral

$$E(P) = \int_{-\infty}^{\infty} \overline{\psi(x)} \, [-i\psi'(x)] \, dx.$$

Exercises

89. We assumed that the particle was moving on the line, so that it is *certain* that its position will be between $-\infty$ and $+\infty$. What is the operator which corresponds to observing whether the position is in this interval? What is the expectation of this observable? What does this imply about the integral

$$\int_{-\infty}^{\infty} |\psi(x)|^2 dx$$

for any state ψ?

90. Take

$\psi'(x) = 0$ if $|x| > 1$,

$\psi'(x) = cx$ if $-1 < x < 1$,

$\psi(-2) = 0$, $\psi(x)$ continuous for all x, where c is a real constant.

Check that $\psi(x) = 0$ for $|x| > 1$. Evaluate $E(P)$. Is its value real?

91. Suppose that $\psi'(x)$ is continuous in $0 \leqslant x \leqslant 1$, and that $\psi(x) = 0$ for $x \leqslant 0$ and $x \geqslant 1$. Prove the identities

(a) $\dfrac{d}{dx} |\psi(x)|^2 = \overline{\psi(x)} \, \psi'(x) + \overline{\psi'(x)}\psi(x)$;

(b) If $\displaystyle\int_0^1 \overline{\psi(x)}\psi'(x)dx = c$, then $c + \bar{c} = 0$.

(c) From (b), show that $E(P)$ is real.

If A is an observable, then when the particle is in state ψ the standard deviation $\sigma(A)$ is given by

$$\sigma(A) = +[\text{var}(A)]^{\frac{1}{2}}$$
$$= \{E[A - E(A)^2]\}^{\frac{1}{2}}.$$

The general form of the Heisenberg uncertainty principle is that if $\hat{A}$, $\hat{B}$, and $\hat{C}$ are the operators corresponding to observables A, B, C such that

$$i[\hat{A},\hat{B}] = \hat{C}$$

then

$$E(C) \leqslant 2\sigma(A)\sigma(B).$$

Hence if $E(C)$ is known and different from 0, the standard deviations cannot both be arbitrarily small. If the observed values of A cluster very closely to its expectation $E(A)$, then the observed values of B will be scattered.

In particular, if $\hat{A} = \hat{P}$, $\hat{B} = \hat{X}$, we find that $\hat{C} = 1$ and $E(C) = 1$ (see exercise 89). Then

$$\sigma(P)\sigma(X) \geqslant \tfrac{1}{2},$$

which is the usual form of the Heisenberg uncertainty principle. As an exercise, you may try various state functions ψ, and see if you can find a state such that $\sigma(P)\sigma(X)$ is less than $1/2$.

In the standard calculus texts, one usually finds chapters on moments, work, and pressure. The physical explanations are usually very cursory and give little real understanding of these concepts. The 'applications' are to the physics of 100 years ago, and may have little connection with the engineering of today. The main purpose of such chapters is usually to clothe exercises on definite integrals in the language of applications, with no effort made to consider real applications.

If we want to give exercises on definite integrals using applied language, we may just as well use the language of modern physics, and also exploit the connections with algebra and probability.

SOLUTIONS TO SELECTED EXERCISES

Chapter 1
2    $2^{200\,000}$

3   For $n > 30$, the numerator acts like zero;
for $n > 300$, the denominator acts like zero.

4   It seems that
$$\lim_{n \to \infty} x_n \to 0.618\,0.$$

7   For $x_0 = 1, y_0 = 2, z_0 = 3, x_n, y_n$ and z_n tend to $1.811\,7$.

8   8 operations

11    $2n$ operations

12   7 operations

17    $1.167\,3$

19    $2.015\,3, -0.062\,5, -1.984\,06$

Chapter 2
5    $256 > 243$

8   10^n

9   6^n

11    $3^3 . 5^4 > 4^7$

12    $1.5 < I(3) < 1.6$

13    $2.66 < I(7) < 2.84$

14    $n = 5$

15 (b) $n = 100$

19 (b) $2.321 < x < 2.322$

22    $I(1) = 0$

25   b'

34    $9/4 < y < 5/2$

35    $8/3 < z < 3$

42    $11/7 < x < 8/5$

Chapter 3

2   1 061 520

4    $x(t+1) = (r+1)x(t)$

7    $|C(1, 0.001)|^2 = 7.398$ $C(2, 0.001) = 7.374$
      $|C(0.5, 0.001)|^2 = 2.717$ $C(1, 0.001) = 2.72$

8 (d) $n = 10^{12}$

13    $c = 0.04$

15 (a) $x_n = (0.95)^n x_0$

19    $x_n = (1 - kh)^n x_0$

20    $C = (1 - kh)^{1/h}$

22    $T = \dfrac{-h \log 2}{\log (1 - kh)}$

23    $h = 0.001, k = 0.069\,31$

28    $P(t) = r^t P(0)$,
      $F(t) = F(0) + td$

29   The ratio is less than 0.99 after 35 years, but never less than 0.98.

37    $\frac{1}{3}t^3 + \frac{1}{2}t^2 + \frac{1}{6}t$

41 (a) $(1 + rh)^2 - A^2 \geqslant 0$

  (c) $C(2h) > C(h)$

43 (c) $a_n \geqslant 0$ for all n

46    $C(h, r) = C(h, 1)^r$
      $E(r) = e^r$

47    $x(t) = x(0)e^{rt}$

50   There is no such point.

57    $x(\theta) = 0$ for $0 \leqslant \theta < 1$

58   (3.10a) again

60   The error is less than 0.05 for $\theta \leqslant 0.27$ and less than 0.005 for $\theta \leqslant 0.095$. When using (3.28) the errors, for the same values of θ, are respectively less than 0.013 and 0.000 48.

69 (a) $x = -(t - 1)^{-1}$

Chapter 4

4   A lot of good it will do you.

7   The keyword is POLITICS.

On p. 106 *in the text* the keyword is CONVERSATION.

10   BOLIVAR

11   BALONEY

13 (a)   The minimum occurs at $x = 5$.

15 (c) $h = 5.2, k = 22.8$

  (d)   The minimum of y is 22.8.

17 (mean of B) = 30 + (mean of C)
18 (mean of B) = 10 · (mean of D)
20 m^2 is smaller.
23 $\langle h \rangle = n^2$

26 (b) $A = 1, B = \dfrac{-2}{N(S)} \sum h(x), C = \dfrac{1}{N(S)} \sum h^2(x)$

(c) $\langle (h - t)^2 \rangle$ is minimal for $t = -\dfrac{B}{2A} = \dfrac{1}{N(S)} \sum h(x) = \langle h \rangle$

27 (b) var$(3h)$ = 9var(h)
(c) $\sigma(h + 5) = \sigma(h)$
29 The smallest value of var(h) is 0, and of $\langle h^2 \rangle$ is 25.

33 (b) The minimum of $\langle (f - th)^2 \rangle$ is obtained for $t = \dfrac{\langle hf \rangle}{\langle h^2 \rangle}$ and is $\langle f^2 \rangle - \dfrac{\langle hf \rangle}{\langle h^2 \rangle}$.

34 (b) $\langle fh \rangle$
(c) 2
(d) $\langle h^2 \rangle - \langle h \rangle^2$
36 (e) $P(X \geqslant 3) = \frac{2}{3}$
37 $P(T$ is even$) = \frac{1}{2}$
40 $P(A'$ and $B) = P(B) - P(A$ and $B)$
41 Yes.
43 $P(Y_1 + Y_2 = 1) = 10/36$
45

k	0	1	2	3	4	5
$P(Z = k)$	0.077 76	0.259 2	0.345 6	0.230 4	0.076 8	0.010 24

47 $E(X) + E(Y) = E(X + Y)$
51 The minimum occurs where $t = \langle x \rangle$.

53 $P(|X - t| \geqslant 10) \leqslant \dfrac{E(Y)}{100}$

55 var(X) = 2.916 7, $\sigma(x)$ = 1.707 8.
   The minimal N is 583 340.
59 $0.141\,593 \geqslant E(\chi_{[0,b)}) \geqslant 0.141\,592$

Chapter 5
15 Choose $x_4 = (x_1 x_2 x_3)^{1/3}$
17 The maximum is 2.
23 $k = v_1/v_2$
29 (a) $F = 1$
(b) $F = 1.5$

33 (b) $Z^* = X^*$

(c) $U^* = Y^*$

34 (a) $\dfrac{\Sigma X_j^* Y_j^*}{\Sigma (X_j^*)^2}$

Additional question: how much is $\Sigma (X_j^*)^2$?

35 Look for straight lines.

37 No.

Chapter 6

2 At the middle point we will have the average of the values now at the outer points.

8 It would take more time to get to the same steady-state temperature distribution.

9 (a) 2001

(c) 261

(d) 365

12 (a) $v_{n+1}/v_n = \left(\dfrac{n}{n+1}\right)^n$

(d) $k = 1, N = 16$

14 $N = 34$

18 $x_{n+1} > \dfrac{3}{2}, \dfrac{\delta_{n+1}}{\delta_n} < \dfrac{1}{4}, N = 9$

19 $\lim x_n = 1.3245$

26 (a) $\lim\limits_{n \to \infty} \log \left(1 + \dfrac{1}{n}\right)^n = 1$

(b) $J(h) = \log (1 + h)^{1/h}$

32 $k = \dfrac{2 - \sqrt{3}}{2 + \sqrt{3}}$

36 It is the equilibrium population.

37 It is true at points where the curve intersects the line $y = x$; zero population; equilibrium.

38 For large $t, x(t)$ approaches $\lambda = \dfrac{R}{c}$.

42 $A = 1 - Rh, B = hc$

57 $p = -K, p = 1/L(r - E)$

58 $\dfrac{dy}{dt} = y^2 (4L - R) + y(\tfrac{1}{2}RB - \tfrac{1}{2}RA - 4BL) + BL$

Chapter 7

1   Yes, it is legitimate.

No, it does not yield the 'expected answer'.

3   No, try $n = 42, 43 \ldots$

9   The motion is periodic, the period is 4, the initial states do not matter.

12   $w(t) + w(t + 1) + w(t + 2) = 0$

25   $v = -7, -17, -27$

      $a = -10$

26 (a) $x(1) = 3, v(3/2) = -12$

 (b) $x(2) = -9, v(5/2) = 24$

 (c) $x(3) = 13, v(7/2) = -36$

      $x(4) = -23, v(9/2) = 56$

29   $v(t + 5h/2) = (h^3 \omega^4 - 2h\omega^2)x(t) + (1 - 3h^2\omega^2 - h^4\omega^4)v(t + h/2)$

36   Ellipses

45   $\dfrac{dX}{dt} - V = \dfrac{dV}{dt} + \omega^2 X = 0$

46   Again, $\dfrac{dX}{dt} - V = \dfrac{dV}{dt} + \omega^2 X = 0$

48   $B = \omega^2$

49   $C = 0$

65   $c = 0$

Chapter 8

5    $u(t,x) = 20x$

8    $V_1 = \frac{1}{2}U_2; V_2 = \frac{1}{2}(U_1 + U_3); V_3 = \frac{1}{2}(U_2 + U_4); V_4 = \frac{1}{2}U_3$

10   $W_1 = \frac{1}{4}(U_1 + U_3); W_2 = \frac{1}{2}(U_2 + \frac{1}{2}U_4); \ldots$

     $|W_1| \leqslant \frac{1}{2}M; |W_2| \leqslant \frac{3}{4}M$

11   $V = (1,2\lambda^2, 4\lambda^3 - 2\lambda + 1, 2\lambda^2 - \frac{1}{2})$

14 (b) $U = (-1,-2,3,-5)$

 (c) $\lambda = \pm \dfrac{(3 \pm \sqrt{5})^{1/2}}{2}$

 (d) $U = (2(V_2 - V_4), 2V_4, 2V_1, 2(V_3 - V_1))$

16 (a) $A(2,3) = (5,-1); A(1,\sqrt{2} - 1) = (\sqrt{2}, 2 - \sqrt{2}); A(\frac{1}{2},\frac{1}{2}) = (1,0)$

 (c) $U = (\frac{1}{2},-\frac{1}{2})$

 (d)   This linear transformation is $\frac{1}{2}A$.

17 (b) $A[B(1,0)] = (-1,1), B[A(1,0)] = (1,-1)$

 (e) $EU = (U_1 + 2U_2 - U_2)$

21 (a) $V = (0,0)$

(b) $V = (0,0)$
(c) $V = (-1,1)$
(e) $V = (1,4)$
(f) $V = (-1,-1)$
(g) $V = (-1,4)$

23 (a) $x = 0, y = -2$
   (b) $x = 0, y = 1$
   (c) $x = -1, y = 0$
   (d) $x = -5, y = 6$

24 (b) $U = (1,i), x = i$
   (c) $U = (1, -1), x = 1$

26 $T^3 U = \frac{1}{8}(2U_2 + U_4, 2U_1 + 3U_3, 3U_2 + 2U_4; U_1 + 2U_3)$

27 $x = 0, y = -3/4, z = 0, t = 1/16$

33 $ad \neq bc$

36 (a) $\lim_{n \to \infty} 2^n/3^n = 0$

39
eigenvalues	eigenvectors
$x = 1$	$(\lambda, \lambda, \lambda) \quad \lambda \neq 0$
$x = 2$	$(\lambda, 0, 0) \quad \lambda \neq 0$

41 No.

54 Take the pair of encoded numbers $\begin{pmatrix} a \\ b \end{pmatrix}$,

find $\begin{pmatrix} 2 & -5 \\ 1 & -2 \end{pmatrix}\begin{pmatrix} a \\ b \end{pmatrix} = \begin{pmatrix} c \\ d \end{pmatrix}$,

and the remainders modulo 26 are the original letters.

55 With $\begin{pmatrix} 3 & -1 \\ -2 & 1 \end{pmatrix}$

57 With $\begin{pmatrix} 3 & -5 \\ -1 & 2 \end{pmatrix}$

62 1, 3, 5, 7, 9, 11, 15, 17, 19, 21, 23, 25

64 No.

65 Decoder: $\begin{pmatrix} 3 & -10 \\ -2 & 1 \end{pmatrix}$

67 There is no left inverse.

69 Yes. $[\hat{D},\hat{X}^2] = 2\hat{X}$

71 $\hat{D}[\hat{X},\hat{J}] = \hat{D}\hat{J}^2 = \hat{J}$

74 $[\Delta, \hat{X}] = \Delta + 1$

75   $[\hat{D}^2, \hat{X}] = \hat{D} + \hat{D}$

80   $\hat{X}$ has both a right and a left inverse.

83   $y = \dfrac{x^2}{2} + \dfrac{3x}{2} + \dfrac{7}{4}$ is a solution to $(\hat{D}^2 - 3\hat{D} + 2)y = x^2 + 6$.

INDEX

absolute value, 109
acceleration, 164, 214
algebra (linear), 234
algebra of linear transformations, 246
algorithms, 137, 172
approximations, 51, 75ff, 85ff, 153ff, 228
arithmetic mean, 109, 144
arithmetic progression, 71
average, 108

base, 253
basic, 7ff
birth rate, 68, 183
boundary condition, 208ff

calculators, 4ff
Cauchy, 177
change of variables, 219
characteristic functions, 113
Chebyshev, 122
Chomsky, 99
circuit, 136
classification, 27ff
codes, 100ff
combinatorics, 28ff
commutative property, 246
commutator, 267
compound event, 115
computation with decimals, 165
conditional probability, 118
conductivity, 236
connected graph, 136
conservation law, 222
convergent sequences, 177
cook-book approach, 89
correlation coefficient, 161
cryptography, 100ff, 258ff
cumulative frequency, 97

decimals, 202
deciphering, 100ff
decisions, 162
de Saussure, 92
difference equations, 57ff, 195, 217, 234
differential equations, 75, 225
differentiation, 265
direct standard alphabet, 100
direct study of differential equations, 79ff, 192ff
direction field, 226
divergence, 172

e, 77, 83, 175
edges, 135
efficiency of programs, 16ff
eigenvectors, 257
eigenvalues, 257
enciphering, 100ff
equilibrium, 96, 164ff
errors, 154, 169
existence theorems, 89
expectation, 120
exponentials, 55, 67, 77

false position, 22
Fermat, 115, 145, 146
fishing, 55
flow-chart, 6, 12, 23
four-dimensional space, 239
frequency, 95, 97, 103, 115
frequency distribution, 103
functional equations, 46ff

games, 36ff, 130ff
geometric mean, 144
geometric progression, 64, 70
graph, 135
growth, 56